REA

ACPL ITEM
DISCARDED

S0-BVN-473

LAKE OPEONGO

Untold Stories of Algonquin Park's Largest Lake

S. BERNARD SHAW

Published by

GSPH GENERAL STORE
PUBLISHING HOUSE

1 Main Street Burnstown, Ontario, Canada K0J 1G0
Telephone (613) 432-7697 or 1-800-465-6072

ISBN 1-896182-82-8
Printed and bound in Canada

Layout and cover design by Derek McEwen

Copyright © S. Bernard Shaw, 1998

Also by the author:
The Opeongo: Dreams, Despair and Deliverance.
Canoe Lake, Algonquin Park: Tom Thomson and Other Mysteries (with Gary Long).

General Store Publishing House
Burnstown, Ontario, Canada

Canadian Cataloguing in Publication Data

Shaw, S. Bernard
    Lake Opeongo : untold stories of Algonquin Park's largest lake

Includes bibliographical references.
ISBN 1-896182-82-8

        1. Opeongo Lake (Ont.)–History. I. Title.

FC3065.A65S53 1998   971.3'147   C98-9008149-0
F1059.A4S53 1998

First Printing 1998

Front Cover photograph:
Looking north from Sproule Bay. Terminus of the Whitney & Opeongo Railway was at the bottom
centre where the Opeongo Road now makes a dogleg turn to follow the shoreline of Costello Creek
to the outfitting store (centre), and the fisheries research laboratory.
Jack Mihell, October, 1997.

Back Cover photograph:
The outfitting store, Sproule Bay, 1995. *APMA AD-L-26*

Contents

Page

Foreword v

Preface vii

Introduction xi

1. Early Days 1

2. The Logging Era 11

3. Dennison Farm 19

4. Opeongo Lodge 29

5. Campers and a Cottager 39

6. Fishing and Research 63

7. Aviation 83

Appendix 1   Opeongo Lake Chronology 97

Appendix 2   Origin of Place Names 99

Bibliography 103

Index 109

Abbreviations in illustration credits and bibliography:

AO   Archives of Ontario
APMA   Algonquin Park Museum Archives
CLS   Crown Lands Surveys, Ministry of Natural Resources, Ontario
DLF   Department of Lands and Forests, Ontario
NMC   National Map Collection
MNR   Ministry of Natural Resources, Ontario
NAC   National Archives of Canada
NAPL   National Air Photo Library
NYPP   New York Press Party

Foreword

OPEONGO – the lake and its very name conjure up the ambivalent soul of Ontario's oldest and most famous park. Huge and wild, with quiet bays and rolling white caps, Algonquin's greatest lake embodies on the one hand the unchanging wilderness imagined by the modern visitors to be the Park's simple, reassuring reality. And yet, that magical name, Opeongo, hints at something else, a past which is unknown and perhaps unknowable, of stories forgotten, events unrecorded, and lives unmarked.

The truth is that Lake Opeongo really does have a rich and colourful human history, one which has marked the lake in many important ways but which nevertheless remains unknown to most present day visitors.

To be sure, a few snippets of the lake's history live on in the memories of guides and outfitters, of Park staff, and of residents of the village of Whitney or other long-time visitors to Opeongo. Some stories, including a few of my own, have found their way into print here and there, but until now, there has been nothing like a comprehensive, unified history of the great lake in the heart of Algonquin Park.

Bernard Shaw has filled that void with this fine book and we owe him our thanks. He also earns our admiration for his tireless tracking down of people, stories, and details, for doggedly burrowing through our Park archives and for amassing a truly remarkable pictorial record of at least the recent history of Lake Opeongo.

If you have ever pulled a big Lake Trout up from the depths of the North Arm, had lunch at the old Dennison farm or slowly paddled into the wind past Bates Island you will have wondered about the people and places of Opeongo's past.

You need wonder no longer. Settle back, turn the pages, and let Bernard Shaw tell you the stories of Algonquin's great and endlessly fascinating Lake Opeongo.

I know you will have a great and rewarding read.

Dan Strickland
Chief Park Naturalist
Algonquin Provincial Park

Preface

IN RESEARCHING THE STORIES of Canoe Lake and the Ottawa & Opeongo Colonization Road, I continually found tantalizing references to events on and around Lake Opeongo. Algonquin Park staff and others advised me that, while many scientific papers had been published on various subjects relevant to the park's largest lake, little was available in "popular" literature. The more I investigated, the more I became convinced that a story existed worthy of recounting. I found major portions of the account tucked away in the park archives and supplemented them with important contributions from many local people and summer residents. Memories being tricky things, however, it was necessary to check against established facts, leading me to the provincial and national archives, and to the Crown Lands Survey office in Peterborough where information had lain dormant for many years. One contact led to another and I accumulated a mass of–sometimes contradictory–information. I am indebted to experts who understand the history of Algonquin Park and had the patience to help me organize the material and accurately record events at Lake Opeongo.

Approximately 150 people assisted me with this book, all giving generously of their time, and some trusting me with treasured photographs and family records. It is with some hesitation that I mention anyone by name because so many contributed. Staff of Algonquin Park provided unstinted help, represented by Jack Mihell who led me to many people and to information I would not have otherwise known existed, Ron Tozer who unravelled records in the park archives, Dan Strickland who guided me through the scientific complexities, and Henry Checko who fed me a great deal of information including clues unveiling the story of John Bates. The flavour of early days in Algonquin Park was obtained from oral interviews conducted with park veterans during the 1960s and '70s and stored in the park archives: I felt that they were speaking to me from the pages. Alan Day in Peterborough unearthed Crown Lands Survey records dormant for many years. Anita George lent me Bernice Lisk's papers that explained the Dennison tragedy. John Griffin provided much of the aviation material. Henry Checko, George Garland, Gary Long, Jack Mihell, Jean Shaw, Dan Strickland, Ron Tozer and Victor Solman all laboured through drafts and made crucial corrections and contributions. I also gratefully acknowledge financial assistance from The Friends of Algonquin Park with the cost of researching this book.

Algonquin Park generates worldwide admiration, appreciation and respect largely because of the people who work there. This book is dedicated to the staff and volunteers, past, present and future, who protect and nurture the park's largest lake–Opeongo, one of the most fascinating in Canada.

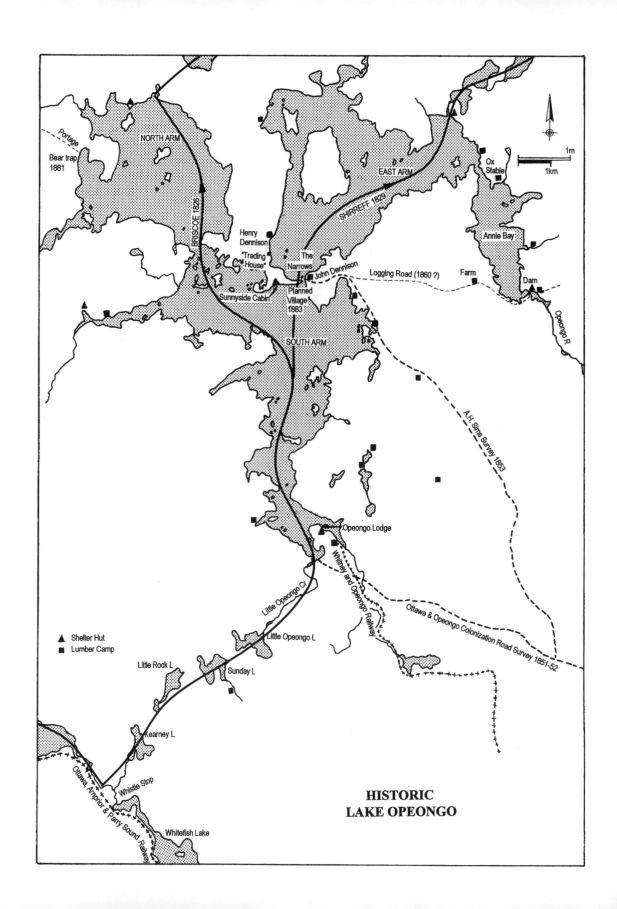

Portage

NORTH ARM

Bear trap
1881

BRISCOE 1826

SHIREFF 1829

EAST ARM

Ox
Stable

Annie Bay

Henry
Dennison

"Trading
House"

The
Narrows

John Dennison

Logging Road (1860 ?)

Farm

Dam

Sunnyside Cabin

Planned
Village
1883

SOUTH ARM

Opeongo R

A.H. Sims Survey 1853

Opeongo Lodge

Little Opeongo Cr

Witlney and Opeongo Railway

Ottawa & Opeongo Colonization Road Survey 1851-52

Little Opeongo L

▲ Shelter Hut
▪ Lumber Camp

Little Rock L

Sunday L

Kearney L

Ottawa Arnprior & Parry Sound Railway

Whistle Stop

Whitefish Lake

1m

1km

**HISTORIC
LAKE OPEONGO**

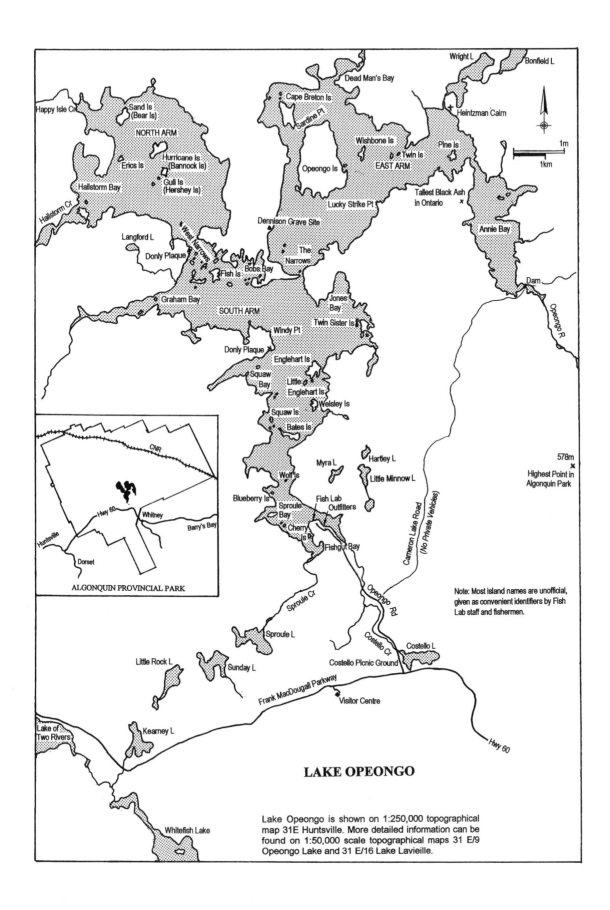

LAKE OPEONGO

Lake Opeongo is shown on 1:250,000 topographical map 31E Huntsville. More detailed information can be found on 1:50,000 scale topographical maps 31 E/9 Opeongo Lake and 31 E/16 Lake Lavieille.

Note: Most island names are unofficial, given as convenient identifiers by Fish Lab staff and fishermen.

ALGONQUIN PROVINCIAL PARK

Happy Isle Cr
Sand Is (Bear Is)
NORTH ARM
Hurricane Is (Bannock Is)
Erics Is
Gull Is (Hershey Is)
Hailstorm Bay
Hailstorm Cr
Langford L
West Narrows
Donly Plaque
Fish Is
Bobs Bay
Graham Bay
SOUTH ARM
Windy Pt
Donly Plaque
Englehart Is
Squaw Bay
Little Englehart Is
Welsley Is
Squaw Is
Bates Is
Myra L
Wolf Is
Blueberry Is
Sproule Bay
Cherry Is
Fishgut Bay
Fish Lab Outfitters
Sproule Cr
Sproule L
Little Rock L
Sunday L

Dead Man's Bay
Cape Breton Is
Sardine Pt
Wishbone Is
Twin Is
Pine Is
Opeongo Is
EAST ARM
Lucky Strike Pt
Dennison Grave Site
The Narrows
Jones Bay
Twin Sister Is
Hartley L
Little Minnow L
Cameron Lake Road (No Private Vehicles)

Wright L
Bonfield L
Heintzman Cairn
Tallest Black Ash in Ontario
Annie Bay
Dam
Opeongo R
1m
1km
578m
Highest Point in Algonquin Park

Opeongo Rd
Costello Cr
Costello L
Costello Picnic Ground
Frank MacDougall Parkway
Visitor Centre
Kearney L
Lake of Two Rivers
Whitefish Lake
Hwy 60

CNR
Hwy 60
Whitney
Barry's Bay
Huntsville
Dorset

Introduction

ON THE FIRST INSPECTION tour of Algonquin Park after it was established in 1893, James Wilson, Superintendent of Queen Victoria Niagara Falls Park, described Lake Opeongo as, "a truly noble expanse . . . when seen in the hazy dawn of an Indian summer morning its beauties make a lasting impression on the mind." It has not changed.

Opeongo. There is a magic ring to the word evoking images of remote northern wilderness and its earliest inhabitants. It is believed to be derived from Algonkian "Ope au wingauk" translated as "sandy at the narrows." This description likely refers to the sandy lake bottom at The Narrows between the South and East Arms, focus of much of the activity recorded in this book.

The promise of free land along The Ottawa & Opeongo Colonization Road lured settlers from the Ottawa River toward Lake

Looking south from the East Arm toward The Narrows (Dennison farm clearing on the right), Bates Island and Sproule Bay. North Arm disappears to the top right. c.1955.
KENNETH AND THELMA WYATT

Ken Wyatt left his job as toolmaker with General Electric in Peterborough to devote more time to his twin passions of flying and photography. He and his wife Thelma often visited Lake Opeongo in their Piper Super Cub.

Lake Opeongo is the largest body of water in Algonquin Park. With a total water area of 5,154 hectares, it extends about 15 kilometres north to south and 14 east to west. The irregular shoreline is 149 kilometres in length (171 if islands are included). Elevation is 403.4 metres above mean sea level. Deepest water, 49.3 metres, is in the centre of the South Arm. A shoreline of modest relief is diminished by raised water levels, but hills south of Annie Bay rise to 578 metres, the highest point in southern Ontario.

Numerous creeks feed Lake Opeongo. The outlet from Annie Bay into the Opeongo River is restrained by a concrete dam capable of a 2.5-metre head of water but normally held at about half this height. The Opeongo River drains south and west to the Madawaska and Ottawa Rivers.

Opeongo. Few of them completed the journey but "Opeongo" records their presence as the name of a range of hills alongside the route and as a street name in many Ottawa Valley towns and villages. Notable among the many businessmen that adopted the name is J.R. Booth, perhaps the most influential logging and transportation entrepreneur in Canadian history. He named his private railway car "Opeongo" and impressed his guests with fine china bearing this name within the Canada Atlantic logo.

Few signs are readily visible today of the momentous events that have taken place on and around Lake Opeongo. For centuries, native people gathered, hunted and fished here, followed by trappers and loggers. Thousands of beaver pelts went to satisfy European fashions. Rafts of squared pine logs were assembled on Lake Opeongo, flushed down the Opeongo, Madawaska, Ottawa and St. Lawrence Rivers to Quebec City and loaded into sailing vessels bound for Britain. Relative peace reigns today as scientists, fishermen, canoeists and campers enjoy the lake.

Arriving at Sproule Bay from Highway 60, the visitor's view up the 15-kilometre reach of Lake Opeongo is restricted by nearby headlands. The reader is taken past these barriers to examine the full expanse of a magnificent lake and to explore its history.

Viewing the uninhabited shoreline beyond Sproule Bay, it is difficult to believe that a century ago ambitious plans were made to establish a village with hotels, farms and mills to serve thousands of travellers expected to travel along a highway from the Ottawa River to Georgian Bay. A logging railway from Whitney evolved into a road serving a large outfitting store that grew from a single log cabin. Pioneer farmer John Dennison was killed here by a bear in 1881, a prelude to more deaths in 1991 on Bates Island, itself a memorial to the only man to build a private cottage on Lake Opeongo. A military camp for boys had a short life in the 1920s and red-carpeted tents were prepared for a visit by Queen Elizabeth in 1959. Artists cultivated their talents here and fishing lures were developed to exploit trout fishing believed by many to be the best in the world. Scientists established one of the world's leading fisheries research laboratories and their work resulted in important contributions to aviation safety during World War Two. There have been many deaths on the wind-lashed lake which has its own ghost story. Now banned, aircraft of many types delivered fishermen and canoeists to Lake Opeongo and its portages leading to the interior of Algonquin Park.

Larger bodies of water usually have "Lake" first, as in Lake Ontario. In this book, deserving dignity is given to Algonquin Park's largest lake by reversing its official name, Opeongo Lake, to Lake Opeongo, as many who know the lake do as a matter of course.

1
Early Days

FOUR HUGE CONTINENTAL GLACIERS advanced over Ontario and then retreated during the period 1,000,000 B.C. to 8,000 B.C. The last ice sheet withdrew from the Algonquin highlands about 11,000 years ago, leaving the land essentially as we see it today. A major glacial meltwater spillway deposited sand through the East Arm and down the Opeongo River. Fish prospered in lakes formed as the ice retreated, and the stark landscape was gradually clothed in green, attracting mammals, reptiles, insects, birds and the first human inhabitants.

Nomadic Indians travelled through the Algonquin highlands hunting and gathering food from about 7,000 B.C. Archeological studies around Lake Opeongo conducted by W.M. Hurley and others in 1970 found quartz scrapers, shards and other material scattered around the shores and islands, indicating aboriginal camps and, possibly, a quartz quarry on a small island off the west shore of the East Arm. A concentration of artifacts found at The Narrows indicates that this was a favourite fishing spot and may have been a settlement. A stone adze was found on the north shore of the North Arm in 1955 by Whitney resident and outfitter Jan Van Baal. Persistent accounts, notably by surveyor James Dickson in 1885, speak of an Indian burial ground near what became the Dennison farm on the East Arm. Loggers favoured the same locations, close to water, as did the Indians, in many cases destroying or covering evidence of earlier occupation with their camps, equipment and modified water levels.

For thousands of years the westerly highway of the native people from the Ottawa River bypassed the barrier of elevated and rugged Algonquin highlands. Experience had proven the best canoe route to be up the Mattawa River to Lake Nipissing and down the French River to

Georgian Bay. This was the route followed by Etienne Brûlé in 1610, Samuel de Champlain in 1615, and the missionaries and fur traders who followed.

Throughout the Algonquin highlands and the rest of the Ottawa River watershed, Indians developed a network of water routes linked by portage trails for hunting, fishing, gathering berries and cutting birch bark. Family groups dispersed during winter seasons to camps in the highlands to improve hunting opportunities. As a matter of mutual survival, families had defined trapping areas generally respected by neighbours. These Indians, named Algonquins, lived relatively peaceful lives until disaster struck in the 1640s. Iroquois marauders from the south arrived on a murderous mission designed to obtain control of the beaver trade. The Iroquois disposed of the beavers and the Algonquins with equal ferocity.

Military explorers followed in the footsteps of the native people, occasionally with their guidance. Early in the 19th century, Royal Engineers were sent to investigate the possibility of building a canal from the Ottawa River to Georgian Bay giving access to the upper Great Lakes. Canada remained apprehensive of further incursions from the United States after the 1812-14 conflict; military minds sought an east-west route more secure than the vulnerable St. Lawrence River. In those days of primitive roads and before the railways had made much impression, canals were seen as the answer to both military and commercial transportation problems. The Canadian government readily committed to the Welland, Rideau and St. Lawrence waterways and thoughts naturally turned to building an east-west ship canal north of the Rideau–as far as possible from the U.S. border. The area between the Ottawa River and Georgian Bay–the "Huron Tract"–contained a labyrinth of rivers, lakes and swamps: surely, these could be connected by a canal.

Most of the military parties explored from the lonely Georgian Bay outpost at Penetanguishene on Gloucester Bay (now Severn Sound). Attracted by the large expanse of Lake Simcoe which promised to set them well on their way east, they normally travelled south of Lake Opeongo. Lieutenant Henry Briscoe, Royal Engineers, and Ensign Durnford, 68th Light Infantry, however, decided in 1826 to follow a route from Georgian Bay which took them right through the Algonquin highlands and across Lake Opeongo to the Petawawa and Ottawa Rivers.[1] Frustrated at being unable to find a knowledgeable guide to assist in his assigned investigation of the Talbot River east of Lake Simcoe, Briscoe reported that he followed the advice of an Iroquois Indian from Lake of Two Mountains and ascended the Muskoka River to its source. Briscoe and Durnford are the first white persons known to have paddled up the Oxtongue River and

1 Henry Briscoe served in the War of 1812 as a junior officer and was subsequently in charge of constructing a section of the Rideau Canal under the command of Lieutenant-Colonel John By. He married while in Canada but returned to England in 1832 (the year the canal was completed) and died in 1838 while serving in Demerara.

through Smoke, Ragged and Porcupine Lakes to the height of land at Bonnechere Lake. Their obvious route then was to follow the flow of the Madawaska until it reached the Ottawa River where Arnprior is now located. In the vicinity of Lake of Two Rivers, however, a second chance meeting with an Indian inspired them to portage northeast to Lake Opeongo which impressed Briscoe as "a very large lake."

Briscoe and his fellow officers turned in negative reports on the practicality of building a ship canal through the Huron Tract. Nevertheless, strong opinions were held–and continued to be held for many years–that one should be built. Entrepreneurs sensed an opportunity to do business. A leader among the proponents was Charles Shirreff, an early settler at Fitzroy Harbour about forty kilometres up the Ottawa River from Bytown (renamed Ottawa in 1854). In 1829, he

Sketch showing modes of communication between Lake Simcoe and the Ottawa. Report of Henry Briscoe, Lt. R.E., 1826. *NMC 2845*

Military officers of the period were not trained as explorers but Briscoe compensated for his lack of precision in map-making by his sense of adventure which took him on a route diverting from his instructions. He was

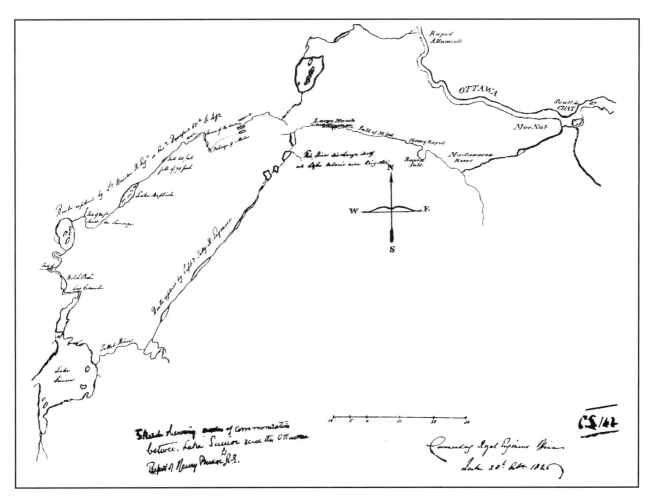

not controlled by measurements of latitude and longitude and his sketch map is barely adequate to identify his route from Lake Simcoe up the Muskoka to Lake Opeongo and the Petawawa and Ottawa Rivers. The route of Lt. Catty up the Talbot and Gull Rivers and down the Madawaska in 1818 is also shown.

dispatched his son Alexander to explore the Huron Tract. The Shirreffs believed that future settlement in this vast region stretching all the way to Georgian Bay depended on canal access–and they wanted a piece of the action.

Shirreff had the benefit of previous explorations and was aware that the traditional route up the Ottawa River to the Mattawa for good reason bypassed the Madawaska, Bonnechere and Petawawa Rivers which lured the inexperienced explorer westwards into their frustrating rapids and shallows. He also knew that the Mattawa had many rapids so he departed from the Ottawa at Deux Rivières, some forty kilometres short of that boisterous river. He struck south through North Depot and Radiant Lakes to the Petawawa River, pausing at the Cedar Lake winter quarters of an Algonquian chief, Constant Pennaissêz, to await advice from his son, absent on a hunting trip. The three-day delay paid dividends for Shirreff as he received "an excellent chart, which delineated the route as far as his hunting bounds extended, nearly to the source of the Nesswabic." (In Shirreff's time, this was the name for the Petawawa River which rises in the Butt Lake area of western Algonquin Park.)

From Cedar Lake, Shirreff made his way through Burntroot to Misty Lake where a fortuitous meeting with an Indian put him on the right path to Canoe Lake, the Oxtongue and Muskoka Rivers and Georgian Bay where he visited Penetanguishene. On his return trip, Shirreff followed the line of Briscoe's exploration of three years before, taking a short cut from Tea Lake through Cache to Lake of Two Rivers, thence into Lake Opeongo and through Wright, Bonfield, Dickson and Lavieille Lakes to the Petawawa and Ottawa Rivers.

Shirreff published his report in the Transactions of the Literary and Historical Society of Quebec (Volume 11, 1831, pages 243-310). He entered the "Peonga Lakes" through the South Arm, proceeding north through The Narrows into the East Arm:

> "The first lake of this name extends northerly in a winding direction one or two miles broad, but of the length I could form no idea, as the route only passes up it about four miles to the outlet, near which there is a trading house belonging to the company, occupied in the hunting season. After gliding down a swift, clear stream for two or three hundred feet, the second Peonga opens to the view, four or five miles in extent, to the northeast, and with a considerable breadth, though its appearance is much diminished by a hardwood island, containing some hundred acres."

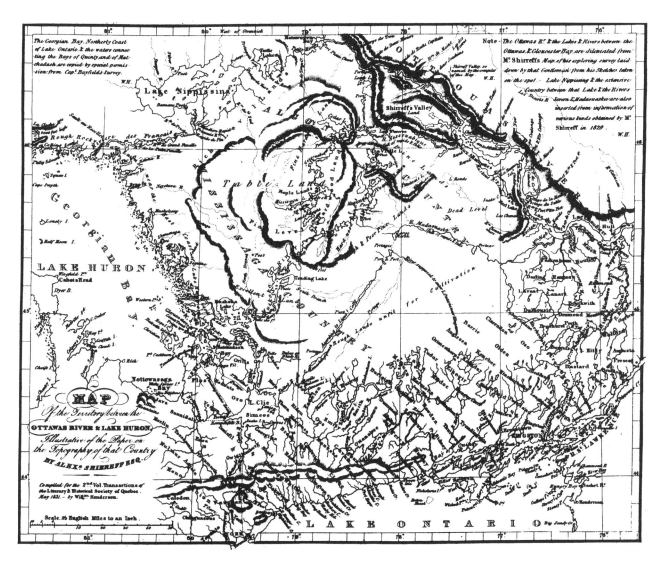

Shirreff's enthusiasm led him to optimistic, even dangerous, projections regarding the suitability of the land for settlement. He described the land north of Opeongo as having the most fertile appearance and to the south of much the same form. Had other reports not tempered his view, the region might have been opened to settlement with disastrous consequences as inexperienced farmers futilely struggled to till the rocks and swamps of the Canadian Shield. In defence of the explorer, it must be noted that settlers viewed all trees as enemies to be conquered. The general view was that conifers grew in poor soil, but a key indicator of fertility was the presence of hardwood trees, numerous in the

William Henderson compiled this map of the Huron Tract in 1831 from information provided by Alexander Shirreff. Although Shirreff had by far the best understanding of the region's geography at the time, many blank spaces remained. The largest, centred approximately on today's

Bancroft, is labelled "Rough Lands unfit for Cultivation." To the north are "Rough & Poor Pine Lands" leading to "Abeunga, or as commonly pronounced, Peonga Lakes," the first known written reference to Lake Opeongo. Early explorers regarded the three arms of Opeongo as different lakes as shown here. Later, they were referred to, collectively, as Great Opeongo Lake.

Henderson attempted to flatter Shirreff by naming the "Fine Land" lying southwest of the Ottawa between Deux Rivières and today's Pembroke, "Shirreff's Valley." A gesture that was apparently not welcomed by Shirreff as a note on the map's margin reads, "For Shirreff's read Maskinongé Valley, being so named in the Notices. A.W."
Map of the Territory between the Ottawa River & Lake Huron . . . By Alexander Shirreff Esq. Compiled by Willm Henderson, 1831.
NMC 2856

2 *Plan and Survey of the River Madawaska–a tributary of the Ottawa–beyond the Surveyed Lands commencing at the southwesterly boundary of the Township of Blithfield under Instructions bearing date 19th January 1847. AO SR11070 No.46.*
3 *Survey Diaries, Field Notes and Reports, Madawaska River, From 5 April 1847 to 29 November 1847. D. McDonell. AO MS924 Reel 13.*

Lake Opeongo area. It was believed that any land capable of nourishing such impressive growth would surely be equally prolific when planted with more mundane crops.

Duncan McDonell conducted a survey of the Madawaska and Opeongo Rivers and Lake Opeongo in 1847.[2] This was a necessary prelude to identifying and selling logging "berths," a major means of generating funds in the days before income and a plethora of other taxes filled the government coffers. McDonell delayed his departure to visit Montreal in an unsuccessful quest for a wider mandate than just a survey of the river. His report described The Narrows as measuring less than a chain (19 metres) wide and his accompanying map indicated a fast flow with a many-feathered arrow. His report reproached the Crown Lands office and nurtured the seeds of disappointment for A.H. Sims and disaster for the Dennisons who followed him:

> "(My report) concentrates on distances and bearings, not agriculture or forestry prospects . . . I have to remark as my instructions confined me almost entirely to the survey of the River, that considerable portions of good lands might have been near the river without me observing it, and which I believe to be the case from accounts I received from individuals whom I considered entitled to credit. I have seen tracts of good land which I believe extend in considerable quantities and that a great part of the Country westward is composed of good land and will admit of extensive settlements, this opinion is grounded not only on what I have seen, and what reason leads me to believe, but it is further grounded on the information I received from a decent intelligent Indian who knows the Country and in which his hunting ground lays who made me a map showing the River with its different Branches and pointing out where the pine ended and where the good land commenced . . ."[3]

The Ottawa and Opeongo Colonization Road was surveyed by Robert Bell in 1851-52 from the Ottawa River near Renfrew to the south end of Lake Opeongo. A branch line north to The Narrows was surveyed in 1853 by A.H. Sims. Free lots were offered to settlers in order to "open up" the region to agriculture and develop an east-west communication link. It was planned to extend the highway all the way to Georgian Bay but realization that the land could not support even a fraction of the forecast eight million settlers ended government funding near Bark Lake, about 40 kilometres short of Opeongo, in the 1880s.

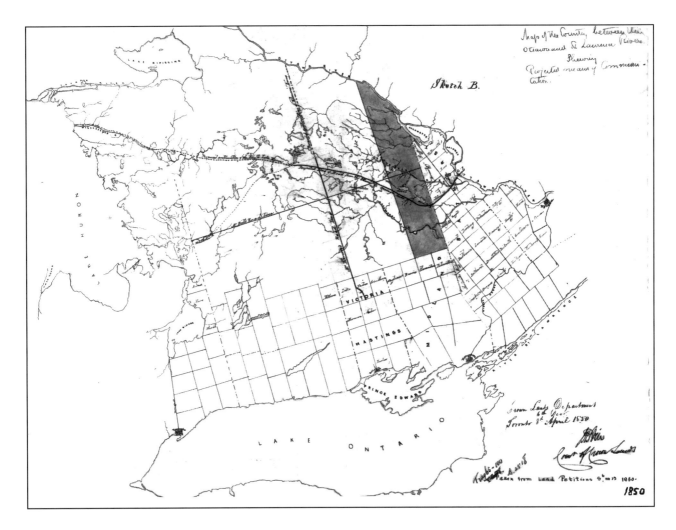

Sims was undoubtedly influenced by the reports of Shirreff and McDonell and by grandiose government colonization plans. Like the Dennisons who followed him in 1871, he probably had visions of a busy commercial centre where the road crossed Lake Opeongo. His field note book illustrated his vision for the village: "By erecting a dam of moderate length a good site for mills could be made at the narrows between the lakes." Other evidence of Sims' intentions to develop the area is shown in his accompanying diary. After being "occupied overnight with Mrs. Sims" in Renfrew, he sent two men with cattle, including an ox, up the Bonnechere on October 9, 1853. Sims and a party of six left the same day, "hindered by head winds and canoe too small." Obtaining a larger canoe on October 13 near

The Government intended to colonize the Huron Tract. This map accompanied a request from the Governor General to build a colonization road from the Ottawa River to Lake Opeongo, shown with a solid line. The dotted line to Georgian Bay shows a proposed continuance.
Map of the Country between Ottawa and St. Lawrence Rivers showing Projected means of Communication. Crown Lands Department Toronto 6th Decr 1850. *NMC 2884*

Government Map of part of Huron and Ottawa Territories 1863-64. Thomas Devine, Department of Crown Lands. AO B-60(1396). Note "Little Opeongo Lake" in Robinson Township, renamed Aylen Lake about 1854 but the old name persisted for many years.

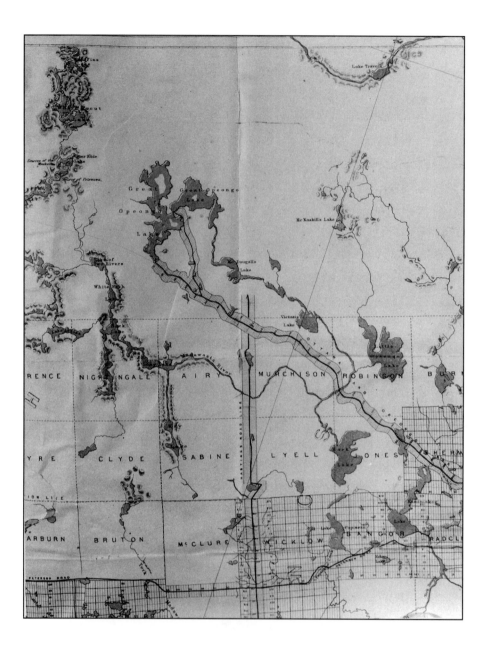

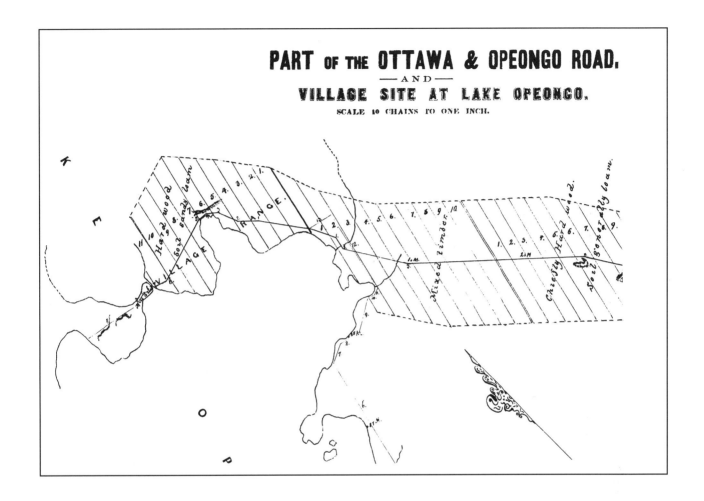

PART of the **OTTAWA & OPEONGO ROAD.**
— A N D —
VILLAGE SITE AT LAKE OPEONGO.
SCALE 10 CHAINS TO ONE INCH.

Calaboga (Calabogie) and in fine warm weather he proceeded through Kanaminiskain (Kamaniskeg), Bark, Victoria and McDougall (Booth) Lakes, arriving at Opeongo on October 22, the cattle two days later. The survey was completed in rain and snow by November 19 and the men retraced their steps to Renfrew where Sims allowed the men five days' pay "on account of the weather" to return home. No mention was made of the fate of the cattle brought to Lake Opeongo but they were possibly left with Alexander McDougall who had a farm on the shore of what is now Booth Lake.

At the terminus of his survey, Sims planted a cedar post five inches square marked "Road line" and deposited underneath it a small glass bottle labelled "Opedildoe." This may have been *Opeleka Liniment*, a favourite cure for, "lame back, rheumatics, swollen joints, headaches,

Detail from Sims' 1853 survey showing a village range on sandy loam soil with hardwood trees and a mill site at The Narrows. A note on his survey shows that he believed a road crossing there would shorten the route westwards by several miles and avoid a spruce swamp southeast of Lake Opeongo. Sims accompanied Bell on the 1851-52 survey and some hostility between them is

bruises, etc." sold by Jas. H. Pullen, Waverly, Nova Scotia. Although the site has been identified by local historians, a forest fire about 1885 probably destroyed any surface evidence. Several investigators have searched without success: the bottle awaits future treasure hunters.

The Algonquin highlands remained essentially undisturbed for thousands of years until the middle of the 19th century when shouts of loggers, the ring of their axes and the crash of fallen trees signalled major changes for Lake Opeongo.

2
The Logging Era

EUROPEAN FASHION DEMANDS for beaver skin headgear brought French trappers and traders into the Algonquin highlands from the Ottawa River in the 17th century. A reminder of their presence was a French axe found by Park Naturalist Grant Tayler at The Narrows about 1965. Britain's victories at Quebec and Montreal in 1759 and 1760 consolidated the gains made by the competing British fur traders although French trappers remained a significant force. The nearest permanent trading post to Opeongo of The Hudson's Bay Company was 90 kilometres east at Golden Lake. Entrepreneurial trapper-traders intercepted furs destined for "The Company" which, of necessity, established its own seasonal outposts. Alexander Shirreff in his 1829 journey noted an unoccupied trading post at The Narrows on Lake Opeongo and assumed it belonged to "the company." No mention of it is found in The Bay's exhaustive records, so it was probably the temporary base of an independent, opportunistic trader.

While a vanguard of trappers, Indian and European, ventured through the Algonquin highlands, an army of loggers made its ponderous advance westward from the Ottawa River, timber cruisers scouting ahead for the best stands of pine. The very existence of logging companies depended on these skilled men, able to find their way through the dense forest and accurately assess not only the number of board feet of lumber that standing trees would yield but also whether the potential return would justify the initial investment. Important factors were the cost of hauling in supplies to sustain men and horses, either by water or by "tote" road, and the merits of establishing farms at the logging sites. Would there be an adequate water supply to flush the logs down river, or would it be necessary to build dams and diversions? Would flumes (wooden troughs) be necessary to bypass rapids? Could these huge expenditures be shared with other loggers? Pivotal to the gamble was the Ottawa River leading to the St. Lawrence and Quebec City where sailing ships would open special hatches in their bows to enable "sticks" to be loaded for the British market. Customers insisted on square timbers that could be efficiently loaded into a vessel and fully utilized at their destination. The waste was left in the forest, often to serve as kindling for forest fires–but that was of little concern with an apparently inexhaustible supply stretching out of sight over the western horizon. (There was later some relaxation to allow shipment of "waney" logs, flattened on four sides but with corners left round, considerably reducing wastage in the forest.) The Madawaska River reached westward from the Ottawa and, eventually, led the timber cruisers some 250 kilometres to Lake Opeongo.

Logging licences of the 19th century recognized the reality of water transportation, awarding cutting rights for a specified number of miles along a river, "following the sinuosities of said river." Five miles back from the river was considered to be the maximum distance to profitably haul a log. (Ironically, logging near the rivers in Algonquin Park is now severely curtailed for environmental and aesthetic reasons.)

Alexander Graham of Renfrew appears to have been the earliest logger at Lake Opeongo, probably before 1860 (Annie Bay, his likely point of entry, was previously named Graham Bay). Many other entrepreneurs followed him.[4] In 1862-63, Alexander and Collin McDonell had timber berths in the northwest corner of the East Arm stretching to Lake Lavieille.[5] To their west, Jeffery Noad and James Skead purchased logging rights.

The McDonells were busy people and one wonders if they were related to Duncan McDonell who undertook the 1847 survey of the Madawaska. Alexander McDonell had a berth on the north side of the Madawaska, "3 miles up and 6 miles back" about 14 miles north of Calabogie Lake. The McDonell's Opeongo berth was transferred to Alexander Fraser about 1871. In 1866, Arthur McArthur transferred to Anderson and Paradis his interests lying south of a line from the North Arm to Johnson's (Happy Isle) Lake.[6] During the cold winter months, freshly-squared pine logs were hauled to the frozen lake by horses. After spring break up, the logs were assembled into booms and towed or warped by "cadge crib" to the Opeongo River at the foot of Annie Bay. A cadge crib carried a horse-powered capstan and solved one of the loggers' major problems–moving logs across lakes. Propelled by oars, it was anchored and a rope or wire cable fed to the log boom which was then winched up

4 Timber Licences RG1 F-1-2 and Timber Limit Applications Books RG1 F-1-4 Vols 2-7. AO.
5 Plan of Petewawa River showing timber licences. A.J. Russel, Department of Crown Lands, Quebec 6 May 1865. AO C-61.
6 Plan of Upper Waters of Madawaska From Eyre and Clyde Townships northerly to Petawawa, 1871. AO R-M(U).

Ralph Bice recalled in *Along The Trail In Algonquin Park* that the Annie Bay and McDougall Lake dams were inoperative at the time of his visit in 1917. The J.R. Booth Company rebuilt both in wood about 1931. Deputy Chief Ranger Aubrey Dunne mentioned in a 1975 interview that in the early '30s, two Booth employees lived in the cabin, seen in the background: "They might have instructions to open the dam at two o'clock in the morning and maybe close it at twelve o'clock noon or something like this you see, in order that the water would reach the point where it was needed at the right working time. I remember there was this French chap and his son, and, of course, he had a lot of spare time on his hands. He had a rock pool built down below the dam and he used to save all the speckled trout that would reach five pounds, keep them alive in there, you know, and when you came

to the cadge crib, and the gruelling process repeated until a water flow fast enough to take over the task was reached. Alternatively, the crib and boom were lashed together and warped up to an anchor carried as far ahead as possible by a small boat. In later years steam-powered "alligators" were used for the same purpose.

In *Algonquin Story*, Audrey Saunders notes that Fraser and McCoshen may have been cutting pine around Opeongo in the 1870s. The notes for her book, retained in the park archives, record that squared timbers were shipped down the Madawaska by McCoshen to Quebec City in 1893. Mort Findlayson, grandson of pioneer settler Captain John Dennison, told her of seeing McCoshen working squared timber through the East Arm some years later, using a cadge crib.

John Rudolphus Booth sent his loggers up the Opeongo River to Annie Bay in the 1880s. Before he pushed the Ottawa Arnprior & Parry Sound Railway through the park in 1896, one of his main supply bases on the Opeongo River was at Depot Lake–hence its name–some 25 kilometres from Annie Bay. Supplies were laboriously hauled to Depot Lake along the Madawaska and Opeongo Rivers, or over a tote road from the Ottawa & Opeongo Colonization Road, itself not much more than a track in places. While the colonization road from Renfrew never did

along there, heck he would give you a five pound trout without blinking an eye."

The dam was raised in 1942 to improve reservoir storage for the new hydroelectric plant at Barrett Chute on the Madawaska River. Rebuilt with concrete in 1955, the dam now also serves as a logging road across the entrance to Opeongo River. The solid bridge will support a fully-loaded truck but damage to the safety railings indicates that at least one found it a tight squeeze.
PHOTO SEPTEMBER 1996

7 *Nigel V. (Nick) Martin was a student at the Fisheries Research Laboratory in 1943, rising to the position of director in the mid-1950s, a post he relinquished in 1963 to compile the results of years of research. He died in 1986 but left a legacy of interest in the history of Lake Opeongo which is drawn upon here in cooperation with his son Chip, a journalist in London, Ontario and other Fish Lab records.*

achieve its final objective, several logging roads reached the lake and gave access to pioneer farms along the way.

Evidence that oxen were used around the northeast side of Annie Bay was found by Nick Martin.[7] Just north of ridges and mounds near the shore indicating the site of a camboose (from the French cambuse - storeroom, also called a shanty), he identified an ox stable with log walls extending more than six feet in height. Martin also explored the site of an old farm, situated on a hillside about one kilometre west of Annie Bay. In the late 1960s it consisted of a roofless house with walls five logs high, a pit near the brow of the hill, and numerous stone piles. The once-cleared section had completely grown in. This farm would have served the first loggers entering Annie Bay from the Madawaska and Opeongo Rivers, and must have been abandoned before the Dennison's arrival in 1871 as they make no mention of it.

Other signs of early logging activity remain around the lake in the shape of root cellar excavations and the remains of buildings and fences. The western portion of Jones Bay in the South Arm is known locally as Warehouse Bay because building remains and numerous artifacts such as axes and horseshoes found there indicates the probable site of a logging supply base.

The virtual freedom of the loggers to change the waterscape of the Algonquin highlands was severely restricted after Algonquin Park was established in 1893. Permission was then required to make any significant change in water flows and levels. Trapping was prohibited in the park, giving rise to continuing conflict between the new park rangers, many of whom had been trappers, and poachers. The importance of Lake Opeongo was soon recognized. James Wilson, Superintendent of Queen Victoria Niagara Falls Park and one of the first team to inspect Algonquin Park, suggested that the park headquarters should be moved from Canoe Lake to a more central location: "Were it not for the difficulty of getting in supplies, Great Opeongo Lake would be an ideal location for this purpose." (Park Headquarters was moved to Cache Lake in 1897.) The 1893 report of the Royal Commission on Forest Reservation and National Park states: "Great Opeongo Lake is the largest and most important sheet of water in that section of the country."

Today's environmentalist looks with incredulous horror at the devastation caused in the last century. Trappers ruthlessly hunted beavers to provide hats for fashionable gentry in Europe. Loggers radically changed water levels with many dams built to facilitate the extraction of logs, drowned uncut trees and destroyed fish habitat. Debris left from logging provided tinder for fires which raged through the area. This must all be considered, however, in the light of attitudes at the time which viewed natural resources as unlimited and free for the taking. A deeply-entrenched feeling was that trees were but an impediment to settlement, agriculture and prosperity. In the days before public assistance was so readily available, the pioneer farmer's very survival depended on his mastery of the forest. He did, in fact, consider trees to be enemies of progress.

Early in 1902, St. Anthony Lumber Company built a railway spur line to Lake Opeongo from the Canada Atlantic Railway at Whitney.[8] The Whitney & Opeongo Railway took in supplies for company logging camps north of the Madawaska and brought out logs, mostly pine, destined for the sawmill in Whitney managed and part-owned by Edwin Canfield Whitney.[9] The weekly *Canada Lumberman* reported that St. Anthony was "putting in" about 40,000,000 feet at the Whitney mill in 1904 and the same quantity the following year.[10] Brian Westhouse records in *Whitney* that J.B. Tudhope, M.P.P. for Orillia, took over Munn Lumber of Haliburton in January 1910 to supply hardwood for his wagon and carriage factory. With partners W.H. Tudhope, A.E. Munn and H.J. Bartlett, he formed Munn Lumber Company Limited and acquired the Whitney mill. The railway, able to bring out heavy hardwood, would have been an attraction. The clear-cut practices of Munn, however, rapidly made the company very unpopular, particularly when trees were felled

8 *The name of Booth's railway was changed from the Ottawa, Arnprior and Parry Sound when consolidated with Canada Atlantic in 1899. It subsequently became Grand Trunk before becoming part of the Canadian National Railway.*

9 *E.C. Whitney, brother of Sir James Pliny Whitney, Premier of Ontario 1905-1914, founded the village of Whitney in 1895. The other mill owners were E.M. Fowler of Chicago and Arthur Hill of Saginaw, Michigan.*

10 *Canada Lumberman, January 27, 1904 and November 30, 1904.*

around park headquarters and the Highland Inn on Cache Lake. In the fall of 1910, the Department of Lands, Forests and Mines bought-out all Munn's interests to 350 square miles in and adjacent to the park for $290,000. The price included removal of all mills and rails in the park. The Dennis Canadian Company obtained the Whitney mill in the winter of 1912-13 together with a 25 year lease from the government for the village of Whitney. It would appear that Munn did not comply with the requirement to remove the Opeongo rails and Dennis used them for some years as park records note their removal in the late-1920s. The line was abandoned by Dennis Canadian in 1926. Highway 60 now follows the railway route from Whitney as far as West Smith Lake, about half-way to Opeongo. The highway now goes direct from there to Costello Lake whereas the railway went via Little McCauley Lake. From Costello, today's Opeongo Road follows the old rail bed to Sproule Bay.

Surveyor James Dickson saw evidence of extensive logging activity by McLachlin Brothers during the 1884-85 winter northwest of Opeongo Lake, where the area was "nearly denuded of the prime timber." He continued, however, to note that no other pine had been taken in Bower Township, except near the lake shore and on the islands ". . . the pine stands thickly on the ground and is of a large and sound quality . . . and will make a valuable timber berth." A 1921 map shows McLachlin

Brothers owning extensive berths north and west of the North Arm bounded by Otterslide, Burnt (Burntroot) and the Petawawa River to Cedar and Trout (Radiant) Lakes.[11] These logs probably sped down the Petawawa River and did not pass through Lake Opeongo.

Brothers Robert J. and James R. Taylor, raised with eight brothers and sisters on the Madawaska River near Combermere at the turn of the century, spent their working lives in logging camps. Jim rose to be a foreman for several lumber companies and Bob was cook for many camps and river drives. In his 1963 unpublished autobiography provided by Hank Legris of Arnprior, Jim Taylor recalls working on the log drive at Lake Opeongo in 1907 for $26 per month. Bob, as "cookee" (cook's helper), received only $24. The brothers must have realized that they were participating in what would become an important period in Canada's history: both left articulate accounts of their careers.

Seven flat cars could be loaded in one hour by two steam-powered hoists at Sproule Bay. A cookery and bunkhouse were also located here along with foreman Jim McKinnon's log office which later became the nucleus of Opeongo Lodge. c.1910.
APMA 3298/R.J. TAYLOR

11 *Drive Route of logs from Timber Limits of McLachlin Bros on Petawawa, Geo. H. Johnson, 1921. AO B-35.*

Alongside a creek on the point of land to the east of Annie Bay is the largest black ash in Ontario—97 feet (29.6m) tall and 30 inches (76cm) diameter at breast height.
PHOTO 1994. JACK MIHELL.

Bob Taylor's story of logging early in the 19th century illuminates the St. Anthony Lumber camp at Sproule Bay.[12] Bob and Jim hopped the returning flat cars at Whitney and had an easy journey to the railway terminus, then called Opeongo Point. Foreman Jim McKinnon charged them with building a dock at Costello Creek to facilitate loading scows to supply bush camps. Evidently not very confident of finding work, they had to return to Whitney for their clothes and it was dark before they bedded down in the bunkhouse. Soon realizing they had unwelcome company, they hastily evacuated to the nearby empty stable. "We got some old hay out of the manger to make a bed. It was a good place to sleep. With the two doors open it was nice and cool and no bed bugs."

With the help of a man and his horse, the Taylor boys drew pine logs and spruce and balsam poles from the bush. By the time McKinnon returned three days later, the dock was finished, evidently to his satisfaction as they were invited to participate in a drive to gather loose logs around Lake Opeongo. "Some had been there for years, and in time of high water had got back in behind tree stumps, rocks, and mud." The salvagers spent most of their time in the water and during two summers, walked "a good part" of the 171-kilometre Lake Opeongo shoreline. Tormented by black flies, they successfully experimented with applications of harness oil found at an abandoned logging camp. "In the mornings and at meal times we rubbed it on. We were in the wind and the hot sun, and when the black harness oil dried, we were black as coal. It took some washing and rubbing to get it off." With typical understatement, Bob remarked that although accommodations and board were "not of the best," not one of the men quit until the work was all done. "There were no unions in the lumber camps. You got so much a month, and if they wanted you to work twelve or fifteen hours a day you did just that and received no extra pay." This work ethic likely gave rise to today's Ottawa Valley lament: "A dollar doesn't do as much for a man as it used to because a man doesn't do as much for a dollar as he used to."

Unrecognized by many visitors, logging in the park continues to be a major factor in the local economy. It is, however, strictly controlled. No sign of logging activity can be seen from Lake Opeongo but, behind a lakeshore barrier untouched for many years, selective logging maintains steady employment. With the cooperation of the McRae Lumber Company a silviculture marking system was first introduced by park staff during the winter of 1968-69 in the area to the west of the South Arm. This system, now used province-wide, ensures that government foresters and commercial loggers "talk" the same language and only trees selected to an established formula are felled.

12 *"Logging in the Valley – 75 Years Ago," Your Forests, Vol. 8, No. 3, Winter 1975, pp 8-12. Ontario Ministry of Natural Resources. Reprinted in Glimpses of Algonquin Park, pp 63-67.*

3

Dennison Farm

Captain John Dennison: a remarkable man with the fortitude to venture into the wilderness to carve a new life when he was more than 70 years old, and be killed a decade later by an enraged bear.
APMA 5720/BERNICE LISK

THE DENNISONS cleared land north of The Narrows about 1871 and were the first white settlers known to have farmed at Lake Opeongo. A logging farm was operating earlier near Annie Bay but no records have been found. The story of John Dennison is a classic example of the well-educated and military-trained middle-class Englishman seeking fame and fortune in the farthest reaches of The Empire.

Born in Penrith, England, about 1799, John Dennison had a commission in the army and helped run the family distillery and brewing business. He married Anne Sanderson of Edinburgh in 1821 and they left for Quebec City in 1831 with three children, Mary, Elizabeth and John Junior. Two more, Henry and Anne, were born in Quebec. He distinguished himself with the Beach River Volunteers during the rebellion of 1837, achieving the rank of Captain by which title he was known for the rest of his life. At the cessation of hostilities, he moved to Montreal where he was demobilized about 1839. While in Montreal, Anne died and was buried at St. Andrews, P.Q.

In 1852, Dennison and his five children moved to Ottawa where he ran a distillery for four years before venturing up the Ottawa River to burgeoning Renfrew County. In reward for his military service he received a grant of land by the Madawaska River where he built a hotel around which grew a settlement named Dennison's Bridge, renamed Combermere in 1865 when Daniel Johnson opened a post office.[13]

Far from retreating from civilization, the Dennisons were hoping to capitalize on the anticipated traffic along the colonization road then being constructed from the Ottawa River toward Lake Opeongo. In the light of knowledge at the time they were perfectly placed to exploit a wide range

13 *Sir Stapleton Cotten, Viscount Combermere (1773-1865) of Combermere Abbey in Cheshire was a senior officer in the Peninsular War and in India. Historian Alan Rayburn suggests that the coincidence of Sir Stapleton's death and changing the settlement's name to Combermere in 1865 could indicate that Captain Dennison served with him and wished to honour his name.*

of commercial opportunities including mills, hotel, farm and toll bridge. Glowing accounts by earlier explorers of the agricultural potential of the area and ambitious government colonization schemes had given them good reasons to leave the security of Combermere.

In the April 1871 census of South Renfrew, Sub-District G, Division 2, Townships of Brudenell, Raglan, Radcliffe and Lyndock, enumerator James Burbridge included John and Henry Dennison as residents of Combermere but their father was not listed. It is evident that the family was at this time well-established at Combermere with a combined acreage of 273, 30 being "improved" and three cultivated to produce 14 bushels of spring wheat, 250 bushels of oats, 25 bushels of peas and 400 bushels of potatoes. Hay was grown on 53 acres. The industrious brothers had two boats and 15 fathoms of net. They reported preserving two barrels of white fish and eight of trout, and were evidently trapping or shooting beaver (8), muskrat (50), mink (30), martens (8), bear (1), and deer and moose (34). They had two horses, two fillies, five cattle and three pigs. Five cattle and swine were slaughtered that year, presumably to supply the logging camps that would be the destinations for most surplus produce. All this was left behind.

Family records state that the Dennisons moved from Combermere in 1870. The absence of Captain John's name from the 1871 census of Combermere could indicate that he went to Lake Opeongo in advance of the remainder of his family, perhaps to survey the territory. Other members may have made prior visits but were listed as residents of Combermere at the time of the census in April 1871.

The Dennisons moved their belongings by canoe along an established trading route from Bark Lake, up the Madawaska, across Victoria Lake and through a chain of lakes to the Opeongo River and the East Arm of Lake Opeongo. Near The Narrows separating the East and South Arms, they cleared land, John on the east side, and Henry on the west which became the main farm site of some 60 acres, known today as the Dennison farm.

Captain Dennison's daughter Elizabeth and her husband John Hudson stayed at Combermere and took over his hotel which, rebuilt several times, is now the Hudson House Restaurant. (Their son John Charles "Jack" Hudson became Reeve of Combermere and famous when he and his crew drowned in the foundering of the paddle-wheeler *Mayflower* in 1912.) By this time, John was married to Elizabeth Bond, and Henry to Ellen Oram.

Mort Findlayson, son of Captain Dennison's daughter Mary, told Audrey Saunders while she was researching *Algonquin Story* in the 1940s that when the Dennisons first arrived at The Narrows they found a small clearing on the west side that they were told the Indians (probably from Golden Lake) had used for a summer camp. It is unlikely that Indians felled the trees but they may have utilized a clearing originated from cutting logs to build a nearby cabin, the remains of which Alexander Shirreff in 1829 thought had been a Hudson's Bay trading post.

During the winter of 1876-77, Henry travelled to Manitoba in an unsuccessful search for better land than offered by the Canadian Shield. During his absence two of his children died, three-year-old Jean and one-year-old Edward. Their mother's distress at the loss of two of her five children can hardly be imagined, but is dramatically expressed by her ghostly image at the Algonquin Park Visitor Centre display: "How much death and misery must we endure?"

Tragedy continued to dog the Dennison's footsteps. At the age of 82, in June 1881, Captain John Dennison had a fatal encounter with a black bear at the portage to Green Lake (now renamed Happy Isle Lake) at the North Arm of Lake Opeongo. As with most tales passed down by word of mouth, various versions exist. This one originated with an account by Bernice Lisk, recorded after she heard it from her grandfather Henry Dennison, who she insisted was "a very truthful, factual man."[14]

In his later years, Captain John normally confined his activities to the garden, leaving more energetic pursuits like trapping and hunting to his sons. Henry set a trap in the North Arm and gave his father explicit instructions not to venture near it. Bernice Lisk wrote, "Apparently, one did not forbid Captain John to do anything." As soon as Henry left to set more traps at the north of Opeongo, the old man rose to the challenge. He enlisted the aid of his eight-year-old grandson John, known as Jackie, and set off in a canoe with his double-barrelled, muzzle-loading shotgun and an axe.[15] Spray from rough water wet the charges in the gun rendering it useless, so the captain took only the axe with him to investigate the trap. Jackie stayed by the canoe and some time later heard his grandfather cry, "Jackie go home!" The frightened boy paddled some ten kilometres home. By this time it was too dark to mount a search but Henry and two other men left at daylight the following day. The trap had been moved and the men spread out in search. The furious bear lunged from the bush at one man, to be shot by a wary Henry who had made sure that his powder was dry. The mangled remains of Captain Dennison were sadly carried home. Death of the patriarch was a terrible blow to the family. An unhappy footnote tells that Jackie died at the age of 18, accidentally shooting himself while hunting moose.

The Dennisons were included in the 1881 census as two households. Captain John (81) is shown as a cooper and part of the household of his eldest son John (44) and his wife Elizabeth (37), and their seven children John (16), Margaret (12), Isabella (9), Charles (7), Mary (5), Agnes (3) and Elizabeth (1). Also included with John were two farm labourers, Robert Hudson (36) and Thomas Hudson (19). (The coincidence in names suggests that the Hudsons travelled with the Dennisons from Combermere where Elizabeth Dennison married John Hudson.) Listed as a separate household was Captain John's younger son Henry (37), his wife Ellen (37), and their five children Annie (14), Henry (10), John (8),

14 *Henry Dennison died in 1936. Mrs. Lisk died May 15, 1991. She passed her records to her relative Mrs. Anita George who made them available. Genealogist Joan Duquette supplemented this with information obtained in collaboration with scientist Nick Martin who took an interest in the Dennison farm from 1943, his first year at the Fisheries Research Laboratory, until his death in 1986.*

15 *Mrs. Lisk gave the gun to an RCMP constable, and the case to Nick Martin who forwarded it to Algonquin Park Museum Archives.*

Bessy (2) and Kitty (four months). Bower Township, in which the Dennisons were located, is in Nipissing District. Only about 60 people, however, lived along the Madawaska and its tributaries west of Renfrew County. It was apparently expedient for enumerator John McMullan to continue up the Madawaska and include them in the Renfrew census as an Unorganised Territory, Sub-District 0-3.

Captain Dennison was buried at his farm. A disintegrating cedar rail fence encloses his grave about 200 metres south of the farm house site and 50 metres from the shoreline. A plaque nearby commemorates the death of two of his grandchildren who were buried beneath two huge white birches about 20 metres north of his grave. In 1996, the site was heavily overgrown and could only be located with the guidance of Ranger Nick Voldock. The birches were dead and decaying.

By 1882 it was clear that ambitious plans for populating the Huron Tract with an agricultural community had foundered on the rocks and reality of the Canadian Shield. Although surveyed all the way to Lake Opeongo, government funding for the colonization road from the Ottawa River dried up when it reached Bark Lake, some 40 kilometres short of its destination. Loggers and farmers did penetrate farther west on their own

Technician Harry Tuvi (left) and Nick Martin at Captain Dennison's grave site, November 1978. The copper "At Rest" marker, below Martin's right hand, has been moved to the black cherry tree growing inside the cedar fence.
APMA 4506/MNR

Left: John and Henry
Dennison photographed
in Ottawa c.1890.
BERNICE LISK/ANITA GEORGE

Right: Ellen (Oram)
Dennison, wife of Henry,
at Aylen Lake c.1910.
BERNICE LISK/ANITA GEORGE

initiative but any hopes the Dennisons may have had for a stopping place, hotel, mills and toll bridge were dashed. Disheartened with their misfortunes, the surviving Dennisons departed. Henry went back to Combermere and, later, to a farm by Aylen Lake (coincidentally, originally named Little Opeongo Lake) where he died at the age of 93 in 1936.

Henry and Ellen's daughter Ellen, born April 8, 1882, married Philip George in 1906. Their daughter Bernice, born in Barry's Bay, was invited by Nick Martin, then Director of the Fisheries Research Laboratory, to the Dennison farm site in 1981, the centenary of Captain John's death.

By 1885, when the survey of Bower Township was completed by James Dickson, the Dennison clearing was being farmed by the Fraser and McCoshen Lumber Company to supply its logging crews. During this period it was called Depot Farm (as were several other farms in the park).

Touched by the tragic experience of the settlers, Mrs. Bernice Lisk (née George), great granddaughter of Captain Dennison, commissioned a memorial to the two youngsters Jean and Edward who died at the farm during the terrible winter of 1876. Mrs. Lisk is shown here at the dedication near her great grandfather's grave in 1982. *APMA 6687/JACK MURDOCK*

Dickson's map and field notes dated May 1885 show that the farm had been extended almost to the southern extent of the peninsula and described the site three years after the Dennisons departed:

> "There is a farm on the peninsula of the Opeongo Lake containing one hundred and twenty-one acres of clearing, well-fenced with large substantial dwelling house, barn, stables and other outbuildings, the property of Messrs Fraser and McCanshon (sic) lumbermen. On it is raised a large quantity of hay, oats, peas, potatoes and garden vegetables for the supply of their lumber camps, besides pasture for a large herd of cattle and a number of horses. This has also been an ancient Indian Fort, the old burying ground being still easily traced."

IN MEMORY OF
CAPTAIN JOHN DENNISON
1799 – 1881
GRANDCHILDREN
JEAN   1874 – 1877
EDWARD   1876 – 1877
REST· IN PEACE

The Dennison farm was temporarily revived in the early 1920s by Colonel Holman James, a retired British Army officer, who established a short-lived boys' military camp. This episode and the subsequent leases to Batchelor and Bates are described in Chapter 5 – Campers and a Cottager.

According to Mort Findlayson, grandson of Captain Dennison, the point of land on which Henry Dennison established his farm was called "Sunnyside." Although Henry's farm faced east, this was possibly because of the point's southern exposure, and to distinguish it from John Dennison's farm on the relatively-shady east side of The Narrows. Timbers from Henry's farm were salvaged to construct a ranger's shelter hut, known as Sunnyside Cabin, on the south shore. The hut has been removed but the foundations remain and the site is now a favourite camp ground with an impressive view down the South Arm.

Joe Avery salvaged some Dennison timbers for construction at Opeongo Lodge and other scavengers were at work. The remains of the Dennison buildings were destroyed. Their location can be detected at the lower-lying ground on the west shore, north of The Narrows.

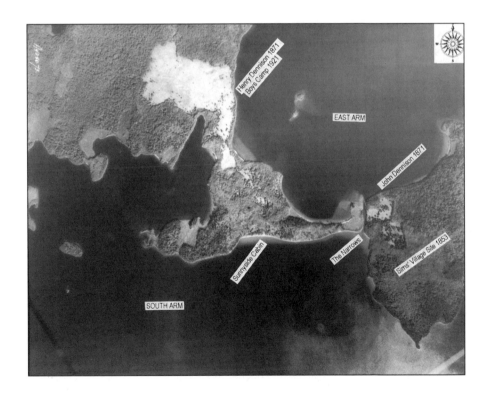

The Dennison farm sites from 11,600 feet August 19, 1931.
NAPL A4280-73

This 1931 photograph of the Sunnyside cabin shows a Rochester, New York, canoe party on an expedition from Camp Pathfinder on Source Lake. Harold "Red" Drake stands in the doorway, Al Makin meditates and Bill Brice smiles in the foreground.
APMA 1427/W.L. BRICE

Robert G. Torrens stands outside the Graham Creek cabin in 1936. There were 115 similar ranger shelters in the park spaced about 25 kilometres (one day's travel) apart. Eleven remain and are kept in good repair for rental by canoe trippers.
APMA 1347/
ROBERT G. TORRENS

The Brice and Torrens families camped on Opeongo and subsequently built cottages on Cache Lake. Robert and Millicent Torrens bequeathed their Algonquin Park records to Bill Brice who made available several photographs for this book.

Cookout with the help of an old stove top at the Dennison farm
APMA 341/MNR/NYPP

The New York Press Party, invited to enjoy and publicize Algonquin Park during the 1930s, contributed many photographs to the park archives.

Fish Lab Technician Greg Betteridge helped locate the Dennison farm site in 1996. He stands alongside asparagus still marking the garden, along with raspberries and rhubarb, a century after they were planted.

4
Opeongo Lodge

A VARIETY OF OUTFITTING STORES operated in the park during its formative years; most were appendages of hotels, lodges and camps. The Parks Division of the Department of Lands and Forests, established in 1954, exercised its authority in 1955 to limit outfitting concessions to one at Canoe Lake, a second at Lake of Two Rivers, and a third at Lake Opeongo.

Today's large Opeongo outfitting store evolved from a single log cabin office at Sproule Bay, the terminus of a logging railway built in 1902 to feed the Whitney sawmill. Following removal of the rails about 1926, the right-of-way offered the only practical means of automobile access to Lake Opeongo and entrepreneurial eyes were evidently cast on the abandoned cabin. Even before the rails were removed, Deputy Minister Cain advised Park Superintendent John Millar that one John Whitton was trying to dispose of a house on Lake Opeongo for which no approval was on file. John Edward Whitton was a butcher in Whitney. (Jack Whitton's daughter, Dr. Charlotte Whitton, was the feisty Ottawa mayor in 1951-56 and 1961-64, and the author of many books including *A Hundred Years A-Fellin'*, a history of Gillies' lumber operations.) Millar identified the house as a peeled log structure originally used as an office by St. Anthony Lumber Company, owned by the department and utilized by rangers as a halfway house. The interested purchaser, Fred Wright, was made aware of the true ownership and attempted to rent the cabin. A Licence of Occupation, however, was issued in June 1925 to Alexander "Sandy" Haggart, who had been using the cabin to store gear for his embryo tourist business.[16] Haggart's rent was $25 per season but delinquent payments resulted in him being placed on a monthly rental of $5 in 1928.

16 *Haggart's name is spelt "Heggart" in some park documents. Unable to read or write, he signed with an "x"*

The old log cabin built by St. Anthony Lumber at the turn of the century was considered to be beyond repair in 1925 and this new ranger cabin was built a few yards to its north. Red pine was cut at the north end of Opeongo and towed to the site by motor boat. The 20' by 24' cabin with a ten-foot "shed" at one end was built by the rangers at a cost of $45 for material and $132 for ranger labour. The outline of the original entrance can be seen on the side wall, the door was moved to the lakeside when the porch was added. This was the second home of Ranger George Heintzman for many years. The fire rangers had a cabin and a hose drying tower nearby, enclosed by a fence which obstructed the park rangers' access to their cabin and was the cause of acrimonious correspondence between the two organizations.
APMA 312/MRS. SWAN.

Because he did maintain a service for tourists, he was reluctantly given a one-year Licence of Occupation for one acre at $20 in September of that year.

Ralph Bice recalls in *Along the Trail In Algonquin Park* that Haggart purchased a Ford car which he used to transport passengers along the railway right-of-way to Lake Opeongo. George Holmberg, who had earlier worked as a blacksmith for J.R. Booth, helped Haggart and the rough ride over the remaining ties inspired him to fabricate two immense iron hooks. Dragged beneath a horse-drawn wagon, they wrenched ties out as he proceeded with his normal transportation duties. The next stage was grading with several ties chained together and dragged behind the wagon until the roadbed was somewhat improved. Left behind were innumerable spikes, discovered over the next decades in the tires of many frustrated motorists, perhaps including Holmberg who later became Deputy Chief Ranger. Despite its shortcomings, this road was the only practical means of entry to the park by automobile until the construction of Highway 60 was started in 1933.

Recognizing a business opportunity in the increasing flow of anglers lured by the exceptional fishing, Haggart named his acquisition "Opeongo Lodge." By 1929, he had improved the access road to the extent that it could be used by visitors–although tales have it that he did little to improve the accommodation at his one-cabin lodge.

A flurry of Department of Lands and Forests correspondence in 1929 resulted from the Forestry Branch's request to reserve the whole point on which Opeongo Lodge was situated for a reforestation zone. This was quickly rejected on the grounds of Haggart's presence and Sproule Bay being the only practical means of access to Lake Opeongo and beyond for a growing number of tourists. The Forestry Branch's reserve was limited to 500 feet at the waterline, centred on the boathouse, by 300 feet deep. (Differences between branches of the Department of Lands and Forests were common until 1942 when they were combined under the same administration.)

In 1930, Haggart purchased Red Row, a bunkhouse built by St. Anthony Lumber in Whitney. With the help of Mr. Fox, the Whitney station agent, he used the material to construct two additional cabins at his lodge. He also struck a deal to take over the old government boathouse, built in 1923, in exchange for materials to construct a replacement. A love-hate relationship existed between the department and Haggart who was often late with his payments and resisted efforts to have him improve the appearance of his unpainted buildings. The park authorities were tolerant, however, because he did maintain the road from Whitney, provide some services for tourists, and guard the department's motor boat stationed on the lake.

Fire rangers and park rangers combined their efforts in 1930 to build a "pavillion" 18 by 20 feet, kitchen 10 by 12 feet and a toilet for the use of a growing number of tourists using a small campground northwest of the lodge. A floating dock 30 by 12 feet was also constructed. The pavilion remained without a roof and the kitchen without a stove until 1931 while jurisdiction responsibilities were sorted out between the Lands Branch and the Forestry Branch of the Department of Lands and Forests. The pavilion, photographed about 1935, was a popular gathering place for dances and cookouts for several years. *W.L. BRICE.*

A major expansion of the campground was planned in 1930 but did not materialize, and the land is now occupied by the Harkness Laboratory of Fisheries Research. The original campground continued in use by fishermen and canoeists until the 1980s and is now part of the outfitting store parking lot.

From 10,000 feet on June 9, 1931, Opeongo Road can be seen running north, following the railway right-of-way alongside sinuous Costello Creek, to Opeongo Lodge. Sandy Haggart's three-cottage lodge is just south of his dock and boat house. The road continues to the campground, pavilion and tourist dock.
NAPL A3423-96

Pilot F/Lt. Sampson and cameraman Sgt. McNamara in RCAF Bellanca Pacemaker "VJ," stationed at Huntsville and part of a nationwide aerial photographic survey, flew over Sproule Bay again on August 19 and photograph NAPL A4280-81 shows a small log boom tied-up by the lodge.

In the fall of 1930, $750 was allocated for "hazard disposal" (i.e., road improvements and brush clearing) along 14 miles of road from Whitney. This was the beginning of "The Dirty Thirties" and half these funds came from the provincial government's special allotment for the relief of unemployment.

In 1936, Joseph E. Avery used the proceeds from the sale of his interest in Mountain Trout House Resort on Kawagama Lake near Dorset to pay Haggart $1,700 for Opeongo Lodge, then consisting of three cabins, one log and two clapboard, and several boats. The department was relieved to have a new tenant as, "he plans to put in a good outfit of boats . . . and (is) better able to deal with the situation than Mr. Haggart who had no funds to put the business in shape." Mary McCormick Pigeon recalled in *Living at Cache Lake* that Haggart started a daily (except Sunday) mail and general delivery service in 1940 between Whitney and Huntsville. Despite his being unable to read or write, she remembers, "This was a great service and we were delighted."

Joe Avery's acquisition had few conveniences in 1936. He extended one cottage as living quarters for his family where they lived the first winter, but in the following years moved to Whitney at the end of the summer season. He was surprised to have the lead party of the newly-established Fisheries Research Laboratory move into the other frame cottage, an arrangement made, and deposit placed, by Dr. Harkness the previous year. Haggart had neglected to pass on the information–and the deposit.

Joe Avery assumed the department's responsibility to maintain a boat on the lake in 1936, perhaps hoping this would favourably influence a decision on his request for a lease on five acres. After a year's delay for a survey to be completed, he received Licence of Occupation No. 4076 in September 1937, retroactive to June, for 1.3 acres. (Leases were normally for 21 years whereas a Licence of Occupation was good for only one year.) He had the frontage requested but the depth was limited to avoid interfering with a proposed road. The cost was $100 and did include permission to operate an outfitting store, but Avery could not plan with confidence past the next renewal date. Apparently disillusioned by the response, he did not immediately proceed with his expansion plans but considered selling the business. In response to an enquiry by a Mr. Brent on behalf of his friend who planned to turn it into "a first-class camp," Superintendent MacDougall provided an inventory in January 1938 which listed three cabins, one large motor boat capable of transporting six to eight passengers, six square-stern boats and approximately 15 canoes.

An additional cabin was built to the south of the lodge in 1938 especially for Mr. Moodie, a regular and favoured customer who warranted his personal Delco generator to supply electricity. A keen fisherman, Moodie regularly left his summer home near Huntsville for a week's fishing in Lake Opeongo. Reputed to have made his fortune in "gold," Mr. Moodie would arrive in his V-12 Lincoln Zephyr that sported a gold hood ornament. He astounded everyone with his ability to consistently catch lake trout of 15 to 20 pounds. Mr. Moodie and chauffeur/boatman Wilf would patiently sit out in the lake all day. His secret was believed to be a large live minnow held in a hooked harness of his invention and lowered close to the lake bottom in deep water. While everyone else was unsuccessfully trolling with artificial lures, Mr. Moodie was happily hauling in the big ones with his live bait. Mr. Moodie was always treated with respect: no one could recall his first name in 1996.

"Lunch on Lake Opeongo."
Ranger Jim Shields in the
car, c.1935.
APMA 95/MNR/NYPP

Probably a "brag" load but
Russel Parks, guiding for
Opeongo Lodge in 1947,
gives a good impression of
the loads carried across
rough portages.
*APMA 2205/CLIFFORD
OLMSTEAD*

Over the years, Joe Avery expanded the accommodation at his fishing lodge and by the 1940s had about 20 men, many from the Golden Lake Reserve, guiding parties through Lake Opeongo and beyond. His children and "hired girl" Ann Byers helped Mrs. Avery with the cleaning and cooking. Lacking refrigeration, Joe and his workers had a busy winter cutting ice and storing it, insulated in sawdust, against the day when successful fishermen would take home their catch. The Opeongo Lodge flier promised "A Sportsman's Paradise" and the front page showed a group of happy anglers admiring 30 trout displayed on an upturned canoe. American Plan rate was $35 per week per person, an extra $5 for "running water."

Road access to Opeongo Lodge had been improved by Sandy Haggart but the section approaching his lodge was frequently flooded. In 1930, the department and the J.R. Booth Company shared the $250 cost of relocating the road to avoid low spots. It remained a very rough trip and became impassable in 1942 when the outlet dam at Annie Bay was raised. A new road was built that year on higher ground from "the landing" (i.e., at the current dog leg approaching Sproule Bay from where the inundated old road can be seen) 2,600 feet to "the hotel owned by Joseph Avery." The total estimated cost was $1,273.20 of which the Hydro Electric Power Commission contributed $700.

Relationships between Joe Avery and the department were as changeable as the Lake Opeongo weather. He continually laboured under the handicap of an annual Licence of Occupation, with no guarantee of renewal from year to year. He was often in arrears with rent and provincial land tax, receiving reminders such as the one in 1945: "remittance to be received immediately if (you) wish to retain rights to occupy the location." The struggling entrepreneur ran into more problems in 1938 when he lent his favourite Savage .250-300 rifle to Basil Sawyer. Sawyer was caught poaching by the rangers and the rifle confiscated. Avery plaintively complained that he would really like to get his rifle back. He didn't, and was again an innocent victim when several lake trout of about seven pounds found dead at Happy Isle were attributed to fishing parties after big fish. As outfitter, Avery was held responsible.

Maintenance of the temperamental Algonquin Park telephone line, strung from posts supported by rock piles in the absence of convenient trees, alongside Costello Creek in the 1930s, was a perpetual problem. Acrimonious correspondence showed that the government had difficulty in obtaining settlement of the Avery account for leased telephone services, a problem compounded by charges for long distance calls placed by Opeongo Lodge customers, long departed. The irritation was not removed until Bell Canada ran in a telephone line all the way to the Fish Lab in 1963, servicing the lodge en route.

Despite making his equipment available to the rangers, Avery was refused permission to use the abandoned J.R. Booth Company cabin at the foot of Annie Bay for "overnighting" his customers because rangers occasionally used it for shelter. In 1946 he did get permission to install a British-American Oil Company glass-front gas pump for retail sales, but was not allowed to supplement his existing sign at Highway 60 advising of this new service. His application in 1950 to extend his 300-foot frontage another 400 feet to the southeast with a 21-year lease was refused on both counts. The park authorities were instituting a policy of owning all outfitting facilities and Avery was advised in September 1952 that, "no further improvements or buildings will be permitted at Opeongo Lodge." This was followed in April 1955 by terse advice from Lands and Forests Minister Clare E. Mapledoram that his licence of occupation would be cancelled at the end of the year, "that date has been set to give you an opportunity to operate for another summer season in the Park, and to give you sufficient time in which to remove any buildings and other structures that you have erected on the land." An appraisal of the buildings was scheduled but Avery was warned that they would become the property of the Crown without obligation for compensation if he did not remove them. The buildings were listed as one log building 35'x40', three log cabins, one frame house 28'x50', seven frame cabins, one Delco house, one boat house, one ice house and one wash house.

Opeongo Lodge c.1957. Guides were accommodated in the left-hand cabin on the far side of the road and the adjacent six cabins were for guests who all shared two strategically-located "honey houses" to the rear. Along the waterfront, from the left, a laundry house; a cabin reserved for favourite guest Mr. Moodie with an ice house at the rear; the main lodge building where the Averys lived upstairs during the summer; and the store with two guest cabins and boat houses in front and a gas pump at the side. Department property started to the right of the store with a small creel census building, dock, ranger cabin, and boat house. The road on the right led to the Fish Lab. Joe Avery's daughter Clover recalls hauling the laundry by hand from the laundry house to clothes lines strung between posts on the hill. Chore boy Clarence Tessier had the job of lighting Coleman lanterns until a generator was installed in the laundry house in 1948, the same year that Jack Brown and Dick Harr built the main lodge.
APMA 6689/KEN WYATT 5835/MICHAEL AVERY

Joe Avery died in 1955 which may have been a reason for the termination notice not being enforced and the lease being renewed to April 1958, at an annual cost of $1,500, in the name of his wife Myra.

The Avery family tradition continued with Joe's son Ken taking over the business in 1958. A major change, however, was in the wind. The Department of Lands and Forests decreed that Avery's business should be limited to outfitting, and the buildings would be owned and maintained by the Crown. $27,000 was paid in compensation for the buildings, fixtures and two motor launches. The name of the Avery business was changed to Opeongo Outfitters, providing canoes, supplies and guides, but no accommodation was available other than for staff.

Joe Avery learned the art of snow-shoe manufacture from his future father-in-law John Boothby in Dorset. Family memories have Mr. Boothby putting Joe to work while he waited for a deliberately tardy Myra. The forced labour paid off and the fame of snow-shoes and paddles made at Sproule Bay for the Averys, their guides and favoured clients soon spread. A business was born that is still operated by the family in Whitney and has been expanded by Joe's grand-nephew Danny to include dog sleighs.

Losing the competition for the Sproule Bay concession in 1977, Ken Avery moved Opeongo Outfitters to its present busy location on Highway 60 just west of Whitney, retaining his popular water taxis on the lake. His sons Michael and Jim now run the business, although Ken can often be seen "keeping an eye on things."

Disaster struck in the spring of 1968 when the store burned just as the Averys were gearing up for the season. Guide Frank Kuiack remembers that the Department of Lands and Forests erected temporary plywood buildings while the store was replaced.
APMA 5409/MNR

The Lake Opeongo outfitting concession passed to brothers Sven and Eric Miglin from 1977 to 1990. Erik (Salty) Sultmanis managed it for them for most of this period under the name "Alquon Ventures Inc." Here (kneeling) he is talking in 1988 to artist and musician Mendelson Joe seated in a water taxi with guide Jim Northcote standing. Sultmanis now manages the Portage Store at Canoe Lake.
E. SULTMANIS

Wendy and Bill Swift took over the Opeongo concession in 1990. They renamed the store Opeongo Algonquin, under the corporate umbrella of their store, Algonquin Outfitters, near Dwight where they also build the famous Swift canoes and kayaks. Bill Swift, an engineer from Rochester, New York, was a young camper at Camp Pathfinder on Source Lake in 1939 when he fell in love with the park. "Tired of corporate slavery at Kodak," he and his wife purchased the Dwight property in 1960. Opeongo Algonquin's manager Jerry Schmanda is kept busy "from ice out to ice in" outfitting and advising canoeists and campers. He claims, "all you need to bring is your toothbrush and clothes." He says that an outfit, including canoe and supplies but not clothing, costs about $45 per person per day. Both Opeongo Algonquin and Opeongo Outfitters offer guided tours from one-day trips to destinations such as Hailstorm Creek in the North Arm, which flows through an immense bog noted for its wildlife, to weeks-long expeditions through the park. Virtually all the 220 canoes available from Opeongo Algonquin alone are in use over the busy summer weekends.

The Sproule Bay outfitters in its various forms has been the funnel for campers, fishermen, scientists, aviators, tourists and the sole cottage-owner who have made their own contributions to the history of Lake Opeongo.

Opeongo Algonquin,
September, 1997.

5

Campers and a Cottager

EVEN BEFORE ALGONQUIN PARK was formally established in 1893, the recreational potential of Lake Opeongo was recognised. Alexander Kirkwood, one the "Fathers of Algonquin" wrote on December 21, 1885, to Commissioner of Crown Lands, Timothy B. Pardee, outlining his ideas and reasons for establishing an "Algonkin Forest and Park:"

> "Seekers for health and pleasure in the summer season may be allowed to lease locations for cottages or tents on the shores of the Great Opeongo Lake, and a site on that lake for a hotel and farm can be offered to public competition at an annual rent."

Access by J.R. Booth's Ottawa Arnprior & Parry Sound Railway was the major reason for cottage development at lakes along its route until the construction of Highway 60 in the 1930s. This railway passed well to the south of Lake Opeongo so there were no cottagers, except for John Bates, to be disturbed by provincial government policies on park development which oscillated from outright promotion in the early days to a decision in 1954 that it be returned as nearly as possible to its natural state.

Logging "roads"–little better than tracks–existed throughout the park but were not passable by automobiles. A logging railway and a subsequent road along its right-of-way provided access for fishermen from Whitney to Sproule Bay. While encouraging fishing, however, authorities were hesitant to open up "interior" lakes such as Opeongo to cottagers. It is also likely that the logging companies resisted cottage development which they would have seen as an impediment to their activities.

No cottages, hotels or farms now exist on Lake Opeongo. There are, however, 176 camp sites around the lake and on its islands. Opeongo also leads canoeists to many more of the additional 1,771 "interior" sites in Algonquin Park. Permits issued for trips starting at Opeongo average about 30,000 every summer.

"Progress" passed by Opeongo and the net result was virtually no development after the loggers left, except for the outfitters and laboratory at Sproule Bay.

It's sheer size was a further deterrent to cottage development on Lake Opeongo. Except on rare, mirror-calm days, it is not for the novice canoeist. Fierce winds can blow up with little prior notice, whipping up white caps, forcing even the experienced to seek shelter ashore. Asked for his opinion of the lake after traversing it as a fire ranger, guide and fisherman for more than forty years, Joe Lavallee immediately responded, "She can get god-damned rough." Metre-high waves can dump the unfortunate canoeist into frigid waters. Notorious shifting winds off Windy Point capsized several canoes of a Pennsylvanian church youth group on July 22, 1970. One of the leaders, a powerful swimmer, righted the canoes and hoisted the boys inside but one panicked, grabbed him, and they both disappeared. A second boy also drowned in the confusion. Police Dive Master Pat Patterson recalls that his recovery boat was anchored in relatively shallow water about 40 feet deep. On his first descent down the anchor rope, he found the bodies of Charles Schnitter, Donald Enzor and Timothy Meadows, two of them still desperately clutching each other.

Even calm weather can be hazardous. On August 8, 1993, a party of Japanese students was canoeing about 15 metres off the shore in the South Arm when 22-year-old Daik Nagashima upset his canoe. He had a life jacket in the canoe but was not wearing it. Despite frantic rescue attempts, he disappeared and was not recovered until the following day by the Ontario Provincial Police diving team.[17]

Early casualties of the lumbering days go unrecorded. Undoubtedly, there were deaths, expected at the time almost as a matter of course. A grave, at best, might have the man's name and the date carved on a board marker by a friend, perhaps with a favourite warning:

> Remember men as you pass by,
> You may be soon the same as I.
> And though you may be stout and brave,
> You might yet meet a watery grave.

Ranger Aubrey Dunne's diary entry for February 10, 1938, contains a laconic entry: "Booth's camp #2 had a truck and a man drowned in East Arm of Opeongo this morning." Over one dozen campers and fishermen have since drowned, several bodies remaining unrecovered.

17 *Occurrence Report, Ontario Provincial Police, Killaloe.*

It was a calm evening in May 1959 when four Belleville men set out from Opeongo Outfitters in two small motor boats with camping and fishing gear. The next morning, a guide found both boats upturned off Windy Point and the floating body of Glen Rogers wearing a life jacket. It was assumed that the boats either collided or one hit a dead-head and the second tipped during a rescue attempt. An intensive underwater search did not find the missing men. On August 31, 1964, the body of Harry Frederick Nickle surfaced.[18] Murray Corneal and Rea Donly were never found.
PHOTO 1996

Lake Opeongo has its own ghost story. While working at the Sproule Bay store in May 1996, Boni Coons was approached by two very agitated campers returning from an overnight at Windy Point. Woken in the small hours by a moaning noise, they claimed to have seen a lady dressed in white wandering through the trees. As they watched, she glided off the rock incorporating Rea Donly's memorial and disappeared into the lake.

The fur trade, logging, the Dennison farm, Jack Whitton's flirtation with fishing in World War One, a fishing lodge which became the outfitting store, and a boys' camp are the only commercial activities to have taken place on Lake Opeongo. From 1908 to 1913, railway companies built the Highland Inn on Cache Lake, Camp Nominigan on Smoke Lake and Camp Minnesing on Burnt Island Lake. Within two decades the number had grown to eight hotels and a miscellany of smaller outfitting stores. None were built on Lake Opeongo, but the Grand Trunk Railway did play a hand in establishing a boys' camp at the Dennison farm, initiating a train of events leading to the only private cottage to be built on the lake.

In August 1920, Holman James, a retired British Army Colonel, followed his introduction to the Department of Lands and Forests by the Grand Trunk Railway System with an application for a "concession" at the Dennison farm site:[19]

18 *Occurrence Report, Ontario Provincial Police, Killaloe.*
19 *CLS 39342 plus information from APMA and Algonquin Park Lease Registry*

Colonel Holman James commissioned a survey of five sites, each about five acres, at the Dennison farm by Speight & van Nostrand, Toronto, completed by Mr. Ward in October 1921. Superintendent Bartlett supported the application, partly because, "it will be the first school camp in the park under a British subject." He resisted James obtaining "Parcel B" because he said it contained a ranger cabin. Lease 2064 for only "Parcel C," which included the Dennison buildings, was approved in January 1922, at an annual rent of $75.

Plan shewing Part Of Lot 27 Con II … etc, Township of Bower. Speight & van Nostrand, Toronto, 14 October, 1921.
CLS 39342

"It is my intention to form a Boy's Military Camp for sons of wealthy Americans to the extent of 250 to 300 boys, next year. This will entail a very great amount of work and expenditure, and will mean an advertisement of extraordinary value to Canada as the boys will be drawn from every part of the United States."

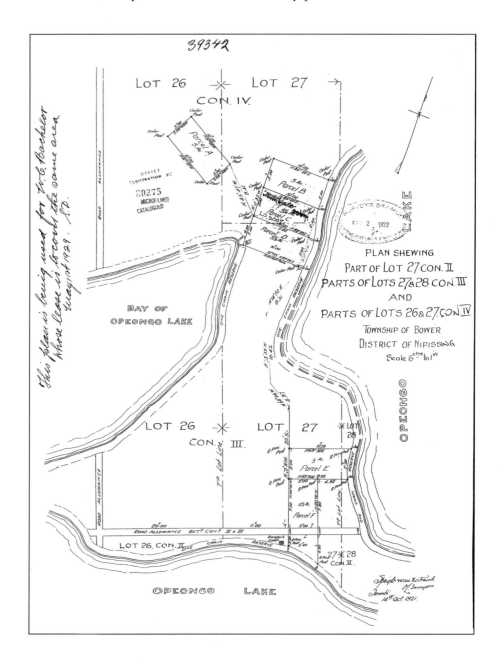

James planned a fleet of 100 canoes and three 25-seater power boats for Camp Opeongo. He promised to bring in two army officers and four senior non-commissioned officers to train the boys who, he claimed, would be of great value in combatting vandalism and forest fires. Correspondence continued for a year, delayed by James' constant travels as an actor (he was also known as Colonel Fred Lindsey) through the United States.

Railway interest in Holman's venture was demonstrated on Canadian National Railway's 1922 map of Algonquin Park which shows "Station for Camp Opeongo (CANOE ROUTE)" at the northwest tip of Whitefish Lake.[20] The railway was actively promoting its lodges in the park and would have seen Camp Opeongo as another business opportunity. Whitefish was probably just a "whistle stop" as there is no record of any buildings there. The 20-kilometre paddle and portage through Kearney, Little Rock, Sunday and Little Opeongo (now Sproule) Lakes to the South Arm and then to the East Arm would have discouraged all but the most enthusiastic boys and staff. This and the difficulties and cost of bringing in equipment and supplies must have contributed to the camp's failure. The Whitney and Opeongo Railway was still operating and offered an easier route to the camp. It would appear, however, that the logging companies objected to the establishment of the camp and Dennis Canadian, who then owned the St. Anthony railway, could have refused Colonel James the use of its facilities.

Early in 1922, J.R. Booth got wind of a "boys summer resort" near his timber berths. He demanded to see the Minister of Lands and Forests and forcefully pressed his concerns about boys getting up to mischief including starting fires. Perhaps because of this intervention, James' lease was not forwarded to the Nipissing Master of Titles until April 1924. In the meantime, James was champing at the bit and complaining that he had already sunk $30,000 into the venture. He did, however, get the camp moving in anticipation of receiving approval. The Brice family camped nearby and Bill Brice was told by his father that James installed a forge and employed a farrier in anticipation of providing his campers with horses.

By the time he received title to the camp, James was engaged with Douglas Fairbanks' company, performing *Don Q* in New York. He requested forgiveness of rent while he was not using the property, and the help of rangers to protect his equipment left without guard at the camp. Neither request was granted. In April, James wrote from California lamenting the loss of his equipment to "thieves and trappers" ("poachers" would have been a more accurate term as trapping was not allowed in the park). He requested that his lease be cancelled. It was, on June 7, 1926.

20 *APMA Q1.4.3*

The Dennison farm
buildings were partially
restored by Colonel
Holman James in 1922-23
as a boys' military camp.
APMA 1424/W.L. BRICE

Acting Park Superin-
tendent Mark Robinson
reported that only two
boys used Camp Opeongo
in 1922 and none in 1923.
He was correct in
concluding that Col. James
would not succeed. A
larger group of Boy Scouts
from Rochester, N.Y. made
use of the buildings in the
early 1930s.
APMA 1415/W.L. BRICE

The Canadian National Railway's Algonquin Park map of that year still identified the Camp Opeongo stop at Whitefish Lake but it was deleted from subsequent issues.[21]

A new player walked on stage in March 1926. Claude LaBarre of Lakewood, Ohio requested that his application for a Camp Ottertrail on Otterslide Lake be transferred to "the vacated site used by Colonel Holman James several years ago for a proposed boys camp which was not successful." The Deputy Minister demurred because he scented other interest for "a large tourist hotel, club house or some extensive operation as such." None of these initiatives proceeded and LaBarre appears to have lost interest. In October 1928 Wilbur Commodore Batchelor, Superintendent of Recreation for the City of Pittsburgh, requested a one-acre lease at the Camp Opeongo site. Park Superintendent J.W. Millar recommended approval but the minister noted that the question of leasing land in that part of the park had still to be resolved and proposed a year-by-year Licence of Occupation. Batchelor reluctantly agreed to L.O.A. 2007, dated May 23, 1929, for all the 5-acre "Parcel C" at $5.00 per acre. He repaired the old log ice house and requested permission to enlarge it, but changed his mind and requested in January 1931 that the L.O.A. be changed to "Parcel D." Objecting to a bureaucracy that required him to terminate one licence before applying for another, and having to pay rent from January 1 when his lease was dated five months later, he terminated his L.O.A. on March 25, 1931. In fact, an earlier application for "Parcel D" had been submitted by John R. Bates and approved, evidently without Batchelor's knowledge, in 1929.

Don McRae of McRae Lumber in Whitney recalls taking about ten teams of horses on flat-bottomed scows to the Dennison farm for summer pasture during the 1920s.
APMA 1423/W.L. BRICE

---

21 *APMA Q1.4.4*

For a time, the Dennison farm was known as "The Colonel's" but all that remained of his enterprise in 1996 was the rotting remains of one of his boats.

John R. Bates was the only person to obtain a cottage lease on Lake Opeongo and visited several times each year for over 30 years.
W.L. BRICE

Algonquin Park was a popular destination for wealthy visitors from the United States in the between-wars years. Among them was pipe-smoking John R. Bates, a Packard automobile dealer from Johnstown, Pennsylvania. A keen fisherman, he was a regular camper on Lake Opeongo during the 1920s. Sandy Haggart took him to the Dennison farm where he established a tent camp, probably to the dismay of his neighbour W.C. Batchelor. Bates evidently had his own ear to the ground and the ear of provincial bureaucrats because he pulled off a very clever move in obtaining permission to build the only private cottage on Lake Opeongo. An application for "Parcel D" was made on his behalf in July 1929 by W.C. Cain, Deputy Minister, Lands Branch, Department of Lands and Forests. Superintendent Millar promptly gave his agreement: "I cannot speak too highly in regard to this party of gentlemen, they are all that could be desired as tenants in Algonquin Park." A Licence of Occupation (i.e., for only one year) was proposed but Bates was awarded Lease 134 of 21 years for "Parcel D." Deputy Minister Cain's letter of July 15, 1929 to Superintendent Millar emphasises, "This is not to be taken as a precedent for accepting or considering any further applications on this lake, and has only been made possible by the fact that there are four or five parcels already surveyed on the ground."

Further evidence of the close relationships between Bates and the park staff is shown in an October 3, 1929 letter from B. McLean who wrote on behalf of himself, John R. Boby, A.W. Leonardson and Bates, inviting Ranger John Boyle to be their guest for ten days in the Adirondack Mountains. He noted the exceptional good time they had in Algonquin Park, thanks to Boyle's courtesy and help, and discretely commented on the value to the park of having him experience the height of the hunting season and witness how "the army of hunters" was handled and camp sites laid out. There is no evidence that Boyle was able to take advantage of this invitation.

Cain steps in again on behalf of Bates on 21 May, 1930. Bates wanted to move nearer to his source of supply in Sproule Bay. Also, he had acquired a 24-foot mahogany runabout, powered by a modified 80-horsepower Packard automobile engine, which he said drew 32" of water and "may be difficult through the channels of the east and west arms." His apparently contradictory desire to be more off the beaten path and preference for isolation could be explained by logging activity around the East Arm and log booms heading through it to the Opeongo River. Undercurrents also suggest that the presence of Batchelor on Parcel C may have been a factor. The truth may well be that this move was part of Bates' master plan to acquire the unsurveyed lot he always coveted. Cain

Two one-acre lots on the northeast tip of "Island A" were surveyed for John R. Bates in September 1930 by F.W. Beatty O.L.S. Bates may have "pulled a fast one" as he obtained these lots on what is now known as Bates Island in exchange for "Parcel D" in the East Arm which he acquired in 1929 only because it had already been surveyed.

DETAIL FROM CLS MAP 86054

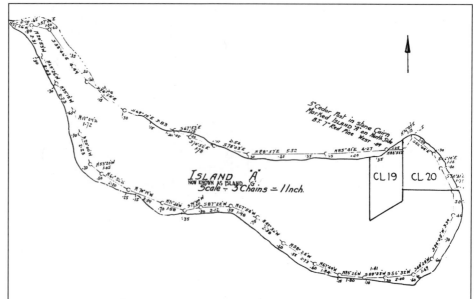

PLAN AND FIELD NOTES

LOCATIONS CL19 AND CL20
ON ISLAND 'A' NOW KNOWN AS ISLAND 'G'
BETWEEN SOUTH ARM AND SPROULE BAY

OPEONGO LAKE

TOWNSHIP OF SPROULE, DISTRICT OF NIPISSING

ALGONQUIN PROVINCIAL PARK

SCALES AS NOTED

NOTE - The field work of this survey was performed between Sept 24th-26th 1930

I hereby certify that the foregoing plan and field notes are correct and are prepared from actual survey made under my personal supervision, and that I was in my own proper person present on the ground during the progress of such survey.

F.W. Beatty
Ontario Land Surveyor

Pembroke Ont. Nov. 15th 1930

In November 1930, Park Superintendent MacDonald observed that Bates had completed construction on Island A in anticipation of obtaining a lease which was, accordingly, backdated to 12 April, 1930, to run for 21 years at 20 dollars per annum. The covering letter from MacDonald reminds Bates of what a good deal he is getting: "I might mention that it is not the present policy of the Department to issue leases for locations on this lake, so your lease will probably be the only one that will be issued for some years." RCAF aerial photograph NAPL A4280/79, taken on August 19, 1931, shows his 48' x 28' frame cottage in place but no dock is visible.
W. L. BRICE

requested that "Parcel D" be exchanged for a lot on "Island A" in the South Arm. A pencilled note on Cain's letter, evidently a draft for a formal response, comments that the boat drew only 20 inches of water but called Bates "a good, straight, honest businessman. No more reliable men come to the Park and they are not fish hogs . . . and he always leaves his boats for rangers' use."

Surveyor "Weary" Beatty determined that Bates' chosen site could not be accommodated on a single one-acre lot, the limit for one leaseholder, as lake frontage would exceed its depth. It was suggested that a family member could own the second one-acre lot but both were eventually registered as one lease to John R. Bates, for reasons not evident until after his death in 1969. He paid for the survey, including the cost of transportation by Sandy Haggart from Whitney. Evidently, Ranger George Heintzman also assisted and was docked two days' park pay (at $4/day) for absence from his regular duties. Correspondence from Superintendent MacDonald approves the ranger's action and recommends that "Weary" reimburse him directly: "Bates is well-fixed and will not mind paying for any extras you may have."

Local residents best remember the Bates' winter visits, probably because they were just about the only ones to relieve the winter solitude. They always arrived in the newest-model Packard. Gordon Palbiski, now

Bates is believed to be one of the few who came out on the right side of the 1929 stock market crash and he retired shortly afterwards, investing some of his wealth in his "estate" on Bates Island. Belle Bates cuts wood in preparation for winter visits when the couple enjoyed the solitude and stocked the ice house, giving her the opportunity to offer the Avery children a rare, and well-remembered, summer treat of homemade ice cream.
W. L. BRICE

The boat house, 32' x 29', included a well-equipped workshop. A Packard engine was installed to haul out his Packard-powered boat, "Vagabond," that Ranger George Heintzman, whose official title was Boat Captain, took great delight in using. The rangers had a key to Bates' workshop which they often utilized to repair their equipment; keeping an eye on the property and delivering mail in return. Bates took great pride in his "North Boy" canoe. All the buildings were of frame construction with cove siding and, according to a 1963 inspection, painted white and in "fairly good condition."
W. L. BRICE

Bates kept in touch with the progress of his investments by a shortwave radio. His wind-driven generator mounted on an observation tower also powered a washing machine. Bates (r), is assisted by Ray Brice, father of Bill Brice.
W. L. BRICE

at Hay Lake, remembers his father Bob and Andy Johnson harnessing up the team of horses in Whitney for the trip to Lake Opeongo. When the road was passable, Peter Bordowitz, who had the general store in Whitney, would pile the Bates and supplies into his jeep and then make weekly visits with more supplies and mail. In the summer months, Bates would leave his car at Avery's store, to the entertainment of the local youths who would sit on alternate sides to admire its automatic-levelling feature—and run down the battery.

For nearly 40 years John Bates did what he loved best—gardening on his island and fly fishing the rivers and lakes around Opeongo, some of which he took delight in keeping "secret." Secret Lake, named by Bates, lies in low-lying terrain west of the North Arm and is difficult to see from ground-level; Bates stocked it with speckled trout and kept this

A large garden was cultivated in the shelter of trees. Mrs. Bates preserved vegetables which lasted well into the fall, despite catering to many visitors. There are frequent and appreciative references in Ranger Aubrey Dunne's 1937-41 diaries to having supper at the Bates'.
W. L. BRICE

information to himself. He was invariably accompanied by Mrs. Bates, remembered by local people as Belle, and often entertained friends, including the Bordowitz' of Whitney, at his cottage and on fishing expeditions. In 1935, a time when wolves were not as popular as they are today, Superintendent MacDougall refused him permission to shoot one, despite an invitation for the superintendent and his wife to visit the Bates cottage where "we will have the fatted calf and perhaps a cocktail or two if you like." MacDougall held fast to the policy that only rangers could hunt wolves and that usually by trapping (in fact, they also shot and snared wolves during this period).

In July 1936, Bates was appalled to receive notice that his lease would not be extended after its term was completed in 1951, "due to plans for Opeongo Lake." He must have exerted influence again because he did receive a further lease on 4 June, 1951 for 21 years and eight-and-one-half months (to bring the annual rent due on January 1) but it included the provisos that no assignment could be made and the lease would revert to the Crown on Bates' death. Bates vehemently but unsuccessfully objected to the new conditions. The new rent for his two acres was $44.75 per annum.

Chip Martin spent his childhood days of the 1950s with his father Nick, Director of the Fisheries Laboratory, and recalls helping his father water the garden during Bates' absence, and enjoying the fresh produce. Mrs. Bates trained a gull, "Charlie," to take food from her hand and it

Left: Guide "Hay Lake" Joe Lavalley frequently assisted John Bates with his fishing and other ventures. Here, he relaxes in the Bates' living room. *W. L. BRICE*

Evidently wishing to have independent means of access to the island, or perhaps tired of finding a perpetually-flat car battery, Bates applied for permission to build a boathouse/garage by Costello Creek (sited approximately at the dog leg in the road approaching today's store). This was initially refused on the grounds that it would set a precedent. However, once again, Bates prevailed and in 1945 received a lease for 0.8 acres at a yearly rent of five dollars plus one percent of the building's assessed value of $100. (The latter amount was later increased to a minimum of $6.) Visitor Millicent Torrens stands alongside her Buick in front of the new garage. *W. L. BRICE*

returned for several years, much to the delight of the youngster. Chip recalls having to assist in rescuing Bates, a big man, when he upset his North Boy canoe powered by a five-horse-power "swing-around" Johnson outboard. Nick Martin and his wife would snowshoe over to the Bates' to play bridge and greatly admired the self-sufficiency of a couple who could live three or four months in their cottage during the winter when both were in their seventies.[22]

Bates claimed compensation for damage to his boathouse, ice house and breakwater in 1958 caused by high waters in Lake Opeongo during the construction of Booth Lake dam. Perhaps this experience cooled his enthusiasm for Lake Opeongo because he enquired in August 1962 if the ministry would be interested in exchanging his property for something similar outside the park. Internal correspondence shows that this proposition was not received favourably. Concern was expressed that such action would set a precedent and it was noted that the lease would automatically terminate in ten years. Bates had already been assured that he would receive the assessed value of improvements made on his lease.

During the mid-1960s the frequency of reminders to Bates of delinquent rents increased. Department letters went unanswered until Mrs. Bordowitz, who corresponded with Belle Bates in Pittsburgh, discovered that John Bates was in a hospital in Johnstown. Sympathetic to Bates' plight, the department took no action until July 15, 1968, when he was formally advised that the leases would be terminated due to nonpayment of rents.

John Bates died in 1969. Lawyers Benson H. Lingle and Charles Bell of Shawville, Pennsylvania enquired in June on behalf of Belle Bates about a new lease in her name. About the same time a letter from lawyer Benjamin Hinchman in Johnstown, representing Mrs. Sarah E. Bates, widow of John Bates, advised the department that she was executrix of Bates' will. This lawyer stated Mrs. Bates' desire to dispose of the cottage, but "she is not familiar with the property and must get additional information before making any decision or attempting to fix a price." The superintendent was asked for information on the value and the department's interest. A July 1970 inventory of contents found little of value except for a Findlay stove and a large oil heater, although it was thought that beds and other furnishings could be utilized in ranger huts. Peter Bordowitz remembers that the large Packard-powered boat had already been sold to Forest Bay Boys Camp on Galeairy Lake.

The apparent existence of a second Mrs. Bates may have compounded the delays in settling accounts. A payment of $5,600 was not made to Mrs. Sarah Bates until September 1972. In the meantime, some

22 *Interview by Roderick MacKay, March 1976. APMA.*

furniture had been removed and the interior of the cottage vandalized to the extent that the remaining contents had no value. It was burned by park staff during the winter of 1972-73.

John D. Robins, professor of English at the University of Toronto, relieved Canada's World War Two gloom in 1943 with his humorous best seller, *The Incomplete Anglers* for which he received a Governor General's Literary Award. He described a canoe fishing trip made with his brother Tom in the mid-1930s. Arriving at Radiant by train from Toronto and North Bay, they canoed and portaged down the Petawawa River and up the White Partridge, through Lavaque, Lavieille, Dickson and Wright Lakes to the East Arm of Lake Opeongo. Robins' lyrical prose provided some of the best descriptions of Algonquin Park's spirit and beauty, and gave a great boost to tourism.

The inexperienced Robins brothers approached Opeongo with some trepidation. They had been told that the lake was animated by a sinister spirit which brooded unsleepingly, malevolently, over the three great arms of that majestic water. Another friend, "whose belief in any kind of spirits is limited to those over a certain per cent proof," told them it was intimidating because of the danger to venturesome canoeists in its great

While John Bates was the only person to build a private cottage on Opeongo, other fortunate individuals did establish a form of residence with tent camps on semi-permanent plywood platforms. Robert G. Torrens had such a camp, typical of many others, on the east shore of the South Arm in the 1930s. *W. L. BRICE*

Ranger Tom Linklater recalls viewing the remains of three tent platforms about 1945 on a small island off the east shore of the South Arm, just south of Windy Point. Eddie Englehart was a regular camper there in pre-World War Two years and the island now, unofficially, bears his name. Ranger Aubrey Dunne complained in his diary on April 20, 1940, that opposite Windy Point he, "... took down tent bottoms that E. Englehart had erected in summer of 1939. He was told to take this down before leaving but did not do so. G. Heintzman and I each worked three hours on this."[23]

23 *A. Dunne diaries, APMA.*

Normally, black bears follow the pattern of most other wild animals: "Leave me alone and I'll leave you alone." Here, Forester Jack Mihell demonstrates a male bear's territorial markings near Annie Bay, 1996.
JEREMY INGLIS

open bays. Swallowing their fears and mistakenly believing there were cottages on the lake, they decided to treat it as "a suburb of Toronto." Their confidence waned, however, when they set eyes upon the vast expanse of water stretching southwards. Robins offered a thought worthy of consideration by all canoeists: "Opeongo was a weather-breeder, a mother of storms, a place of sudden changes of air pressure. It was a malignant spirit, cunning, treacherous, the custodian of all the hatred against the encroachments of white men for three hundred years." Tremulously, they pushed off under a glowering sky and with great relief eventually paddled up to the outfitting store in Sproule Bay where their car awaited.

Probably due to cuddly Teddy Bears and cosy bedtime stories of Goldilocks and Winnie The Pooh, even adult bears are considered to be "cute." They are not. John Robins cultivated the harmless image with an account of his encounter with a black bear. Meeting on a portage, Robins was imprisoned under the canoe he was carrying. In his shaking hands, it oscillated backwards and forwards, almost lifting him in the air as the stern hit the ground. He was exhilarated by meeting the "inoffensive creature" that showed only curiosity at the antics of the brothers seeking to disentangle themselves from their canoe.

Black bears were once hunted without mercy for their thick pelts, and grease used to soften leather and to lubricate wooden axles. Even rangers would shoot them in the park's early days for no apparent reason other than their proclivity to raid shelter cabins for food. A guiding principle of the park master plan now recognizes their prior rights of occupancy, subject to reasonable consideration being given to visitor safety and property. "Garbage" bears mooching around campgrounds certainly have to be treated with caution but are viewed as nuisances by park staff who mark them with ear tags and transport them far away when they push acceptable behavioural limits. A "predacious" bear that, unprovoked, will attack a human is a much more serious matter. Bates Island was the scene of one such event in October 1991, when an adult couple was killed by a 140-kilogram male bear. (The only known prior similar attack in Algonquin Park was in 1978 when three boys died near Lone Creek at the eastern edge of the park.)

In the Bates Island attack, Carola Frehe and Raymond Jakubauskas had apparently started to prepare a meal and a tray of ground beef was still untouched five days after they died. A reconstruction of the event indicated that Frehe was attacked first. Jakubauskas apparently tried to drive off the bear with an oar, found broken at the scene. An empty gasoline can indicated that in desperation he may have tried to douse the bear with gasoline and set it on fire. Both victims were killed by a single blow to the head. The bear was shot and the subsequent autopsy found nothing physical to explain its highly-abnormal behaviour. It was unknown to park authorities and was not tagged so was unlikely to have been accustomed to feeding at camp sites. The summer of 1991 was a particularly good year for a variety of bear foods. It seems that black bears, unlike grizzlies, rarely kill people just because of intrusions into their space or a perceived threat to their cubs. So, if the couple did nothing to incite the bear, it was not looking for hand-outs, and it was not diseased or particularly hungry, what brought on the attack? Chief Park Naturalist Dan Strickland reasoned that, just as an aberrant human may commit a bizarre act, this bear was an off-the-end-of-the-scale predacious individual that took mercifully rare action.

It was a remarkable coincidence that eleven-year-old Nicholas Aikins was mauled by a bear at Happy Isle Creek in the North Arm in July, 1997 only a short distance from where Captain John Dennison was killed by a trapped and enraged bear in 1881. The circumstances, however, were very different for eight boys and two counsellors on the second night of a week-long canoe trip from Camp Arowhon on Teepee Lake. With shouts and kicks, counsellors Michelle Hayes and Mike Hildebrand had twice driven off a bear that had walked on their tent and they were prepared when it returned about two a.m. This time, the bear ripped open one of the tents containing four boys and dragged out young Nicholas. The resulting uproar apparently caused the bear to drop Nicholas. Hildebrand wore a head-mounted lamp and dazzled the bear while belabouring it with a paddle held in both hands. Under constant attack and probably blinded and confused, the bear retreated up a tree permitting a rapid evacuation to the Opeongo outfitting store. Dan Strickland gives credit to Hildebrand for his brave action and for noting the bear's yellow ear-tag which enabled the rangers to shoot the right bear of two in the area, its identity subsequently confirmed by fresh bruises. Nicholas' wounds were repaired in a Montreal hospital. Autopsy of the 142-kilogram male bear found nothing abnormal, indicating again that once-in-a-while, for no apparent reason, a male black bear will decide to attack a human.

Following the 1991 tragedy, Ministry of Natural Resources biologist Mike Wilton, with the help of technician Jeremy Inglis, initiated a comprehensive study of adult male black bear behaviour. Trapped animals had a radio transmitter attached to their collars, released and tracked. Sixteen bears were "collared" up to April 1997, with a maximum of seven "active" at one time because of the time involved in monitoring and continuing attrition due to equipment failures and accidents. Much has

Tracking bears, Jeremy Inglis found himself standing directly over the den of male black bear "Wilfred," weighed-in at 176-kilograms, who had overwintered on the west shore of Graham Bay in the South Arm for the previous two years. Detecting movement below, Inglis hastily departed.
JACK MIHELL, FEBRUARY 28, 1997

Bears love to nibble on nuts at the beech tree-covered hills east of Annie Bay, building up fat reserves for the winter. The chance accumulation in crotches of broken branches resulting from their feasts gives rise to tall tales of them building nests before flying south for the winter. Bears are also attracted to this area in the spring when suckers are plentiful in the numerous small streams.
PHOTO SEPTEMBER 1996.

been learned about the lives and movements of Algonquin's male bears but there is still no conclusive explanation of why an individual male bear may choose to prey on humans. While the danger certainly exists, the odds of encountering a predacious bear are almost nil. Since the 1978 attack, the Bates Island couple are the only fatalities in some ten million visits to the park where about two thousand bears roam at will.

Beside the logging road at Annie Bay dam is a typical sunny patch of sand favoured by snapping turtles for incubating their eggs. Every summer, females lay thousands of eggs, many of which are uncovered and eaten by foraging foxes, raccoons and skunks. A few tiny turtles do survive three-month's incubation and manage to reach the water where they have few enemies, some "snappers" living as long as humans.

Ranger George Heintzman worked out of the Sproule Bay cabin and lived there most of the year while his family stayed in Whitney. His daughter Gwen Giles remembers going out with him on patrol, "My dad always carried a compass and I remember different times he used it, once in particular when he was checking on a burnt-out forest fire. The fire was out on the surface but the roots were smouldering for many months and he kept returning to the same spot." Heintzman could be relied upon to help anyone in trouble and was highly regarded by all the lake residents and visitors. He was the voice of authority on the lake and Nick Martin remembers him hauling in poachers.[24] One example he gives concerns Indian Joe Tennisco who used to venture up the Madawaska to Opeongo in his birch-bark canoe during the early 1950s. Heintzman caught him red-handed at Redrock Lake about two a.m. "This Indian was travelling all over the bush in the middle of the night. Good Lord. I tried travelling in the bush in the night: it's just about hopeless." The impounded canoe was stored at the Fisheries Research boat house for several years before going to Kandalore Canoe Museum, near Dorset. Joe Tennisco was highly amused by the whole affair and built himself another canoe.

To many, George Heintzman was–and still is–the quintessential park ranger. On September 23, 1943, *The Globe and Mail* used him as model for a day in the life of a forest ranger. After a hectic day, he sat by the camp stove and summed up his duties: "I manage to keep busy. I issue travel permits for the park, sell fishing licences, watch for poachers, trap wolves,

Key men at Lake Opeongo in 1945. Left to right, standing, Ranger George Heintzman, Unknown. Seated, Chief Ranger Tom McCormick, Fire Ranger Simon Lynch, Ranger George Holmberg.
MILLIE DARRAUGH

24 *Interview by Rory MacKay at Maple, Ontario, 1976. APMA*

Seen here fishing on Lake Opeongo c.1960, George Holmberg's high-spirited son Hartley worked at the Fisheries Research Laboratory. He died as the result of a motorcycle accident on Opeongo Road in July 1965. Hartley Lake, east of Sproule Bay, is named in his memory.
APMA 1831/DARRAUGH

cut trails, watch the telephone lines, fight fires and try to teach youngsters and adults alike how to act in the bush." He could have added guide, mentor and confidant to premiers, cabinet ministers, senior military officers, journalists and ordinary tourists–with all of whom he was equally at ease. In 1943, Heintzman was paid $125/month plus a cost-of-living bonus of $23.75. George Holmberg, by this time Deputy Chief Ranger, received the same pay but, in recognition of his rank, it was listed as an annual salary.

George Holmberg was a man of many talents. He represented Canada at the prestigious shooting competition at Bisley in England; he was a skilled mechanic and transformed a motorcycle into a snowmobile on which he used to zip around Lake Opeongo. He was on the park staff for 44 years from 1922 and preceded George Heintzman on the Opeongo beat where he travelled in winter by dog sled.[25]

After his appointment as Park Superintendent in 1931, Frank A. MacDougall made strenuous efforts to influence officials and politicians in favour of the park. Heintzman and his colleague Aubrey Dunne, for example, would organize fishing trips for ministry officials and other "VIPs." They would usually operate from a sandy beach in the East Arm, even building an ice house there to store their catch. In his diary for June

25 *Personal communication with Tom Linklater.*

19, 1939, Dunne reports that in one day he and Heintzman took in tiles for a well, dug ten feet deep, installed the tile and backfilled. They also painted the cabin floor and installed paper on the roof to catch dropping gum, and blazed a new portage to Wright Lake 1,200 feet northwest of the original trail close to the cabin. MacDougall ordered the addition of a 14-by 16-foot sleeping cabin in 1942 and persuaded the rangers to paint it both inside and out. In 1945 he ordered grass seed for the site.

Occasionally, Heintzman was reduced to human size. Nick Martin said he was sensitive about his limited formal schooling and saw the white lab coats worn by the scientists as flaunting their higher levels of education. One day in the early '50s, about 20 labbers donned white coats and boarded Heintzman's inboard Firefly, lined up "like a bunch of Vikings." Called out of his cabin, Heintzman saw his favourite boat, which they did not have permission to use, and a sea of white coats. "Well, he was going to arrest the lot of us. He was furious. That was just too many white coats at one time."[26]

With some reluctance, Malcolm Williams (of the Williams Wabler family) recounted the story of his uncle, Doctor Tournay Williams, and the bear. "Doc" regularly took his large family and their friends to his favourite beach on the East Arm, a site he called "tent city." His medical services were in frequent demand so he habitually carried his bag and supplies for which he found use when the camp site was harassed by a black bear one summer. The bear enjoyed a juicy steak laced with morphine and retired peacefully. Rangers were fond of whiling away wet days with "tall tales" and this sounds suspiciously like one of the collection. It continues . . . Heintzman on patrol found the bear fast asleep by the trail and encouraged him to depart with a swift kick to the rear end. Nursing a massive hangover, the bear did not take kindly to this cavalier treatment and the ranger departed at great speed. A popular theory at the time was that any animal not behaving normally could be rabid (in fact, there is no evidence that bears get rabies), so Heintzman called up support with guns. Fortunately, the misjudged bear had departed by the time they arrived.

The most famous people planning to visit Lake Opeongo never arrived. Queen Elizabeth, over a month into a taxing tour, had troubles with a "stomach upset," so she and Prince Philip decided to relax at Batterwood House, Governor General Vincent Massey's Port Hope home, instead of camping in Algonquin Park for the weekend of July 25-26, 1959. (Prince Andrew was born on February 19, 1960.) Premier Leslie Frost planned to attend the party but withdrew when the Queen cancelled.

George Heintzman died on October 4, 1955. The list of attendees at his funeral and those sending messages of sympathy read like a roll call of famous Algonquin park names of the era. In recognition of his service, Deputy Minister Frank MacDougall supplied a plaque and ordered construction of a 16-feet-high cairn on the sandy beach at the East Arm. The estimated cost of $300 was exceeded by $298.66 and had to be explained in detail by Ranger Joe Tate, "straw boss" for the job. Mounting bolts of the plaque deteriorated and it is now stored in the park archives, awaiting replacement. It reads "IN MEMORY OF G.W. HEINTZMAN - ALGONQUIN PARK RANGER - 1938-1955." *PHOTO 1996*

26 *Interview by Ronald Pittaway, October 29, 1975. APMA.*

Park personnel had been busy for weeks preparing the site on the East Arm's sandy beach. RCAF Canso flying boats and provincial and military Otters flew in supplies and communications gear. Red carpet eased the discomfort of walking from the waterline to the royal encampment consisting of about 18 tents on plywood platforms. Logs were salvaged from the old Dennison farm to build an ice house loaded with imported ice and sawdust. A Captain Dubé brought in a large motor boat from Parry Sound to ferry campers and to keep guard at The Narrows.

Retired fire ranger Joe Lavallee remembers being called out at midnight to shepherd an Eaton's truck loaded with bedding down a logging road to the site. He laughs today at being refused a night's rest by the security guards, despite his cargo, on the grounds they had nowhere for him to sleep. Ranger George Pearson recalled there being many cases of "gall darn whisky," and someone he refused to identify sneaking out a bottle for the rangers to share.[27] Another ranger, Charlie Levean, was not too pleased at being one of a group issued with white shirts and bow ties and "volunteered" to serve as waiters. His wife Marguerite was momentarily expecting their first child so both were relieved when the royals decided not to sample the privations of summer camping and he was able to return home for the birth of Terry on July 18.

All the preparation was not, however, wasted. *The Toronto Star* reported from Whitney on Monday, July 27, 1959 that royal party "hangers-on" and some RCAF personnel from Trenton, about 60 in all, had a royal time the previous weekend at the East Arm. Three Lands and Forests float planes and an RCAF "Rescue" Otter were on hand to transport fishing parties in beautiful weather to five nearby lakes where aluminum boats had been cached. George Pearson surmised that they were pretty good "booze hounds" and the ice was not wasted, but the party spirit was dampened somewhat when a demonstrating firefighting aircraft dropped its load a little too close. One lasting benefit, Lavallee recalls, is that plywood used at the camp for tent platforms was utilized in building the East Gate of Algonquin Park.

27 *Interview with May and George Pearson at Bancroft by Rory McKay, c.1975*

6

Fishing and Research

LOONS, NOW A FAVOURITE SYMBOL of the wilderness, were viewed in a very different light a hundred years ago. In his report following the 1893 inspection of Algonquin Park, James Wilson noted that all the lakes were well-stocked with salmon, lake trout and grey trout, but continued: "Large numbers of the young of these fish are annually destroyed by gulls and loons, and it might be advisable to consider the propriety of waging war upon the latter, as neither bird is of much commercial value, and their depredations largely outweigh other considerations." Peter Thomson, first superintendent of Algonquin Park, was of the same opinion, adding, "Bears and foxes should also be destroyed without mercy."

In the early days, fishing was the predominant park activity. Officials knew that spawning beds and feeding areas could be damaged by the flooding of lakes, a fairly common event to facilitate the movement of logs. Restrictions were quickly placed on when and for how long logging companies could dam the rivers, considerably ameliorating the damage caused by frequent changes in water levels. As always, however, there are two sides to the story. Higher waters flooding stony beaches have, in some places, added spawning areas attractive to trout and bass.

The only commercial fishing in Algonquin Park was in Lake Opeongo. Prompted by a meat shortage during the first World War, a licence was awarded to Whitney butcher Jack Whitton to net whitefish and lake trout. Although it is known that he operated from 1916 to 1918, no records have been found of the size or quantity of fish caught. The catch must, however, have been significant to justify his costs, even if he was able to hitch a ride on the railway to Whitney. Ottelyn Addison estimates in *Early Days in Algonquin Park* that it would not have been

Superb fishing was the main attraction to Lake Opeongo. Guide Ned Bowers displays a 32-pound lake trout c.1935. *APMA 106/MNR/NYPP.*

Old-timers boasted of catching 50-pounders but the largest lake trout on record at Opeongo Algonquin store is a 34-pounder caught by 15-year-old Jason Hoover from Bancroft in 1991.

worth Whitton's time to gut and pack any fish weighing less than two pounds.

Algonquin Park, a vast laboratory for research and study of the natural sciences, was, and still is, a magnet attracting scientists. A Forestry Camp was established by the University of Toronto as early as 1908 at the site of an old lumber camp at Burnt (now Burntroot) Lake. From 1924 to 1935 a semi-permanent camp was operated at Achray, then a station on the Canadian Northern Railway. Records of logging operations and tree growth during this period have been important factors in understanding and managing the development of the park. Other early studies by biologists investigated the habits of beavers, hares, ruffed grouse, small mammals and molluscs.

Starting in 1919, the Ontario Department of Lands and Forests undertook studies of many waters in the province, including Lake Ontario, Lake Nipissing, Lake Nipigon, the Trent Canal and trout streams in southern Ontario. In 1929 the first expedition was made into Algonquin Park by W.E. Ricker and F.P. Ide who studied lake trout ecology at Wolf and Ragged Lakes. Soon after his appointment as park superintendent in 1931, Frank MacDougall campaigned for the employment of professional foresters and biologists and, in particular, for a park-wide fisheries study directed at the specific needs of the park and its increasing flow of visitors. He was strongly supported by University of Toronto Professor J.R. Dymond who subsequently played a key role in park scientific activities. MacDougall eventually obtained permission to invite Dr. William John Knox Harkness, also a professor at the University of Toronto, to identify a site for a permanent fisheries research laboratory. Initially focussing on Cache Lake, Harkness decided on Lake Opeongo in 1935 because it was the largest body of water in the park, centrally located and conveniently close to other important lakes and rivers. Sproule Bay offered an appropriate site for the laboratory and it was accessible by primitive road to Sandy Haggart's Opeongo Lodge, already the focus of fishing activities on Opeongo and adjacent lakes.

The Lake Opeongo Fisheries Research Laboratory–"The Fish Lab"–opened for business in the spring of 1936. The staff "made do" in temporary quarters until they could move into the new laboratory buildings under construction at Sproule Bay. Licence of Occupation Number 4513 was issued to the University of Toronto in December, 1938 for 5.87 acres in Sproule Bay at a nominal cost of one dollar per year. The university and the Department of Lands and Forests agreed in 1946 on cooperative use of the facilities but a formal agreement was not signed until 1954. Under the terms of this contract, the Department of Lands

and Forests was responsible for the operation and maintenance of the laboratory and for the provision of facilities for university workers and their programs. An advisory committee consisting of three representatives from the university and three from the government was formed as part of the agreement.

A major objective of the Fish Lab was to develop an understanding of fish and their habitats and to ensure the most effective utilization of what was sure to be a very popular resource. Angling demand was anticipated to increase at a great rate as the park became accessible by automobile with the completion of Highway 60 in 1936. Under the direction of Dr. Harkness, fisheries research was continued in Algonquin Park. (He directed similar research in Quetico, Sibley and Lake Superior Provincial Parks.

Dick Miller was a student in the spring of 1936 when he accompanied Professor Fred Fry and Mrs. Irene Fry to establish a summer headquarters for the Fisheries Research Laboratory. He described the trail alongside Costello Creek as becoming progressively more primitive and petering out at what had been the terminus of the old railway line (the dogleg in today's road). Ahead, the southern extremity of Lake Opeongo appeared to be an enormous expanse of muskeg. Night was approaching and the temperature dropping but Miller was overwhelmed by the smell of spring in the North, "a compound of breaking buds, very early blossoms, and rich, wet earth —

"The leafless trees provided frequent glimpses of little lakes, still partly icebound; white-tailed deer bounded off the road as we rounded each bend, then stood, ears erect, staring at us from the still inadequate cover of the bare bush. A black bear with two cubs made such leisurely way for us that Fred had to stop and change gears. Several moose lifted their heads, as yet uncrowned, and solemnly stared at us as we passed. Even a lynx sprang out on our left and crossed lightly in front of us. I spent several summers in the park but never saw another lynx."

Richard B. Miller describes his first trip alongside Costello Creek to the Fish Lab in 1936. *A Cool Curving World.*

invigorating and yet, somehow, strangely nostalgic and subduing." Waiting for them, tied up at a small dock, was the Fish Lab's transport, an ancient, battered open launch with a Ford Model T engine. Squishing around in the swamp, they transferred gear to the boat and pushed out into a wet tangle of brush. Exiting from the narrow, winding Costello Creek which looked black, mysterious, and sinister in the fading twilight, the advance party emerged into Sproule Bay. They erected a tent alongside the ranger's cabin, then occupied by a fire ranger and his assistant.

Their camp was not quite paradise: vast angry clouds of mosquitoes and blackflies had descended upon them. "Soon rivulets of blood ran down behind our ears, into our collars, and from eyebrows into eyes. Fred's pregnant wife, a city girl on her first trip into the woods, must have suffered most but didn't complain." Miller recounts that this courageous lady looked more unhappy when, after they were bedded down for the night, Fred insisted in reading aloud a chapter of *The Origin of Species*.

Professor Ray Langford and students Bob Martin and Bill Kennedy joined the team early in the summer of 1936. Dr. William Harkness, his wife Martha, and another student, Victor Solman, arrived on July 5. The last member of the initial team, student Ken Doan, joined in mid-August, about the time the roof was put on the log laboratory building. Pioneer "labbers," as they called themselves, worked and slept where they could, utilizing the Opeongo Lodge cabin supplemented by tents erected on the hillside at the rear the lodge. Some moved into a highway construction camp at what is now the Costello Lake Picnic Ground when the engineers moved out in 1937. Gradually moving into new quarters as they were constructed, all were accommodated at the Sproule Bay facility by 1940.

All the labbers either had or obtained their master's degrees or doctorates and were destined to hold important scientific positions in Canada: Victor Solman had a varied career which included meteorology with the RAF, Chief Limnologist with the Department of Northern Affairs and Natural Resources, and international recognition as an expert on bird strike damage to aircraft. Richard Miller became Head of the Department of Zoology, University of Alberta and a member of Canada's National Research Council. Ray Langford was Professor and Associate Chairman, Department of Zoology, University of Toronto and an Assistant Director of the Fish Lab. Kenneth Doan became Chief Biologist for the Province of Manitoba. William Kennedy and Robert Martin became senior scientists with the Fisheries Research Board of Canada. Fred Fry, professor

Clearing the land and construction consumed many hours of the early days. Being one of the original promoters, Park Superintendent Frank MacDougall had a keen interest in the laboratory and flew in at least once a week to check on progress. On one memorable occasion, he tersely queried the work of Mick Koulas, a local workman who had half the logs in place for the lab. building. "Where's the blueprint?" he demanded. Koulas, looking worried, "You only showed it to me." An embarrassed MacDougall dug in his briefcase and found that he had

in the Department of Zoology at the University of Toronto and a Director of the Fish Lab, received the British Empire Medal for his wartime service. J.R. Dymond was Head of the Department of Zoology at the University of Toronto, initiated the park's interpretive program and assisted the Fish Lab from his cottage on Smoke Lake. F.P. Ide was a professor, Department of Zoology, University of Toronto. William Harkness was a professor of limnology at the University of Toronto and became Chief, Fish and Wildlife Branch, Ontario Department of Lands and Forests: the Fish Lab was renamed in his honour.[28]

Top: The main laboratory building under construction, July 26, 1936.
APMA 6692/VICTOR SOLMAN

Bottom: The completed laboratory in 1937. Mrs. Martha Harkness is sitting on the steps. It is still in use. Victor Solman, recalls helping build a frame cottage for Dr. Langford in 1937 and a tribute to his workmanship is the fact that it also still exists, located to the south of the main laboratory. Several staff members, including Solman and his wife Ruth, and Jean and Bill Sprules, used it as a honeymoon cottage.
APMA 3442/DAVE WAINMAN

28 *The Harkness Laboratory of Fisheries Research. N.V. Martin, Department of Lands and Forests, 1968*

indeed carried it away. Always a gentleman, he apologised profusely. A quick inspection showed that Koulas had from memory built exactly as specified.

Cooking was on wood stoves, with a Coleman stove in reserve, and light was from Coleman lanterns. The privy was "up the hill." All the staff

A second "Honeymoon Cottage," its name attesting to its continuing occasional use, was built on the hillside to the north of the complex in 1942.
PHOTO 1996

and the wives had TABT (Typhoid, paratyphoid A and B, and Tetanus) vaccinations before leaving Toronto; they drank untreated water from the lake for the first few weeks but the number of tourists increased dramatically and Solman's diary records that on July 10, 1936 they decided to drink only boiled water–"mostly in the form of tea." Hot water came from a kettle, or a boiler on the end of the wood stove; Solman remembers being entertained by the sight of Ray Langford taking a hot bath in a wash tub on top of the stove.

Three boats were transferred from the fish laboratory at Lake Nipissing, closed with the opening of the new laboratory at Opeongo. The "fleet" is seen here in 1936. One was a heavy pointer, known familiarly as "Umpum," because of the distinctive noise made by its Ford Model T engine. This was the boat the first group had been relieved to see waiting for them at Costello Creek. The other boats were a, 16-foot clinker-built skiff with a two-cylinder, four horsepower Johnson outboard and an 18-foot shallow-draft "speedboat" with a two-cylinder, 14 horsepower Evinrude outboard.
APMA 6693/VICTOR SOLMAN

In addition to their own boats, the labbers had the use of two 16-foot canvas-covered canoes on loan from the park. The skiff was carried on a trailer for serial observations of water chemistry and temperature, fish food and fish population on Brewer, Eos, Two Rivers and Cache Lakes. Canoes were used on Redrock and other lakes. When the weather was not too rough, the speedboat was favoured for the long trips necessary on Opeongo. Solman, one of the junior staff, had to make himself useful as a "go-fer," assisting the more senior scientists. Brought up at his father's garage in Toronto and "handy," he naturally inherited the job of boat maintenance and recalls that all his skills were needed to start the two-opposing-cylinders, early-model outboards.

A steady stream of scientists made their way to the new fisheries laboratory at Sproule Bay. Particularly welcome in 1937 was Oxford University Professor Charles Elton, considered by many to be the world's foremost animal ecologist. Elton had completed an analysis of the Hudson Bay's fur trading records, then in London, that demonstrated a relatively stable beaver population over the previous two centuries while most other animals experienced significant peaks and lows. His theories regarding the influence of food and disease on animal life cycles were of great interest to the Fish Lab staff but Elton had never seen a beaver. University of Toronto professors took advantage of his attendance at a meeting of the Royal Society of Canada in Ottawa to invite him, and Dr. Alan Coventry,

Fish labbers enjoy a rare moment of relaxation with visitors at Costello Lake in 1937. Left to right, Jack Buchanan, Edith Huntsman, Elizabeth Huntsman, Bill Kennedy, Dr. Huntsman, Mrs. Huntsman, Dr. Huntsman's sister, Jon Arnason, and Donald Robb (forestry student). Victor Solman is in the cab of the University of Toronto's 1926 Dodge ("Lyon's Tea" could be seen under the dark paint). Just visible in front of the truck cab is their 1926 seven-passenger Studebaker sedan with bright blue upholstery. The cans were used for transporting fish.
APMA 1962/SOLMAN/COVENTRY

Jon Arnason, a student from Iceland, built a xylophone consisting of 15 whisky bottles salvaged from a road-builders' dump and suspended from the vehicle maintenance ramp. He tuned them with varying water levels and entertained both labbers and visitors in 1937.
APMA 6694/SOLMAN

Much of the information in this section was drawn from the meticulous diaries maintained by Victor Solman and from photographs taken in 1936 and 1937 by him and by University of Toronto Professor Alan Coventry.

to the Fish Lab. There, Victor Solman recalls, he and Dick Miller broke a hole in a beaver dam conveniently nearby in Costello Creek, and built the visitors a comfortable observation post from where they watched the beavers repair their dam. Several labbers subsequently studied under Professor Elton at Oxford.

In June 1937, Solman had a task typical of the times. Used to repairing square holes punctured in the balloon tires of the Studebaker by spikes remaining from the railway, he had a more challenging task when a rear spring broke about two kilometres short of Cache Lake on his return trip from Toronto with the Harknesses. They had to make-do at the Costello camp until Ken Doan returned with a new spring nine days later. Of course, Solman had the job of replacing the broken spring. While the seven-seat Studebaker sedan, purchased for $85, was commodious and versatile, it had no "guts" so Solman had his father ship a new head gasket, installed it and had the pleasure of experiencing the car's vastly improved performance.

Solman was certainly kept busy. Along with Bob Martin he was assigned to assist Irene Fry with the cooking. The students prepared breakfast but Mrs. Fry did "the serious cooking." Solman's eyes twinkle when he recalls her butterscotch muffins. Neither Costello nor Sproule Bay camps had electricity and the two juniors had to cut, split and pile firewood and haul water from the lake between their limnological tasks. Rangers George Heintzman and Aubrey Dunne would cut ice during the winter and store it in a small house north of the laboratory. (In his diary for January 6, 1939, Dunne reports pulling in 274 blocks 16 inches thick, "675 will fill the ice house.") The go-fers had to haul it daily to the cookhouse, either by wheelbarrow on land or by skiff on water.

Bill Sprules sits in a metre-cube mesh cage in 1937 counting insect inhabitants while they devour him.
APMA 6695/SOLMAN

Kennedy and Martin examined the numbers, location, distribution and growth rates of several fish species. Solman, under Langford's guidance, undertook extensive studies of fish food in lakes and streams. Fry and Doan concentrated on the physiology of fish which involved blood chemistry and its relation to oxygen requirements, acquiring knowledge of value to their wartime development of aircrew "G-Suits" and oxygen systems.

Enlisted to help Ken Doan catch bass, Solman, (a meteorologist during World War Two) had an experience that may help explain some of the many drownings in Lake Opeongo. The weather was fine with just a few fluffy clouds and no sign of thunder conditions. Trolling from a canoe in one metre of water at the north end of the South Arm, near the infamous Windy Point, they were surprised by a swishing noise in the trees accompanied by wild gyrations of the treetops. "A miniature tornado came out of the woods about two hundred yards away, crossed the lake and disappeared over the hill." They considered themselves very lucky to be off the twister's path as they were certain it would have upset their canoe.

The Avery's lakeshore cabin was inadequate for the whole crew so on the hillside behind they built platforms for 10' by 12' tents where the parking lots are now located. Miller wrote:

"These were lovely places. They were high enough to get the breeze, which kept them cooler and freer of flies, and also provided a beautiful view of the lake. Here it was only a half-mile wide, and the opposite shore was high and heavily timbered. To sit in one's tent doorway and, with binoculars, watch the forest creatures coming to water in the evening was a continual source of pleasure. Deer we always saw, and they became commonplace; less often there were bears and, rarely, a moose."

The students were kept hard at work but, as might be expected with a group of high-spirited young men, there were a few lighter moments. One evening, the students patiently manoeuvred a porcupine into the bedroom of Victor Solman, renowned for his obsessive neatness. The spectacle of an alarmed Solman standing on his bed with his boots on was considered adequate reward for their dangerous labour. Of course, this affront could not pass without retribution. It was not long before Doan had to dismantle his comfortable bunk, full of sweet, dry meadow grass, in order to remove a four-hundred-pound boulder.

Before joining the Fish Lab, Bill Kennedy had gained much valuable experience fishing commercially with his family on Lake Erie. He applied time-proven techniques to catch fish in trap nets anchored to stakes driven into the lake bottom. Among the many practical innovations he introduced that helped with the scientific research was his "pile driver." Utilizing timbers cut during construction of the laboratory, he built a raft with a central hole surmounted by an A-frame hoist. A heavy weight was raised and dropped to pound in stakes at the trap site.

Record books of the scientists working at the Fish Lab are stored in the Ministry of Natural Resources library in Peterborough. They exhibit both the excitement and the frustrations of the early days. Dr. Harkness wrote in his 1937 diary of Dr. Fry setting out from Toronto early on April 13 to rendezvous at Whitby, east of Toronto, with a truck bringing three cans of cisco fry from the Glenora hatchery that he was to take north to Costello Lake. While waiting at Whitby, he enjoyed breakfast and repaired a puncture. The truck driver fortunately failed to make the rendezvous. Harkness, in Toronto, received a telegram from Superintendent MacDougall advising that the road to Costello was under two feet of soft snow and impassable. Harkness and Dr. Ide drove in pursuit of Fry and found him still waiting patiently at Whitby. The Glenora truck, at last discovered, was sent to plant the cisco fry at Port Credit. Returning post-haste to Toronto, the three scientists rushed over to Eaton's store to get a refund on Fry's load of perishable food supplies. That evening, Harkness completed his day by speaking to about 60 members of the Toronto Aquarium Society. Harkness was evidently shocked by the news on May 6, 1937 as almost the sole diversion from the steady iteration of fisheries research business in his diary reads: "The Zeppelin *Hindenburg* exploded and burned with great loss of life as it landed at Lakehurst today."

The record books also shed some light on the value of a dollar in 1937. Negotiations with pilot Tom Higgins to fly from his base at Limberlost to pick up seven cans, each containing about 1,000 fish and weighing 75 pounds, at the Pembroke hatchery and deliver them to Snake Creek resolved on a price of 70 dollars for the 300-mile trip. In September, Harkness comments, with a note of surprise, that Joe Avery charged visiting Dr. Huntsman and his family $1.50 per person for bed and breakfast at Opeongo Lodge.

Superintendent MacDougall pulled every string he could to put park management on a professional and scientific basis. Unable to obtain approval to hire a professional biologist, he resorted to the subterfuge of appointing Dr. Duncan MacLulich as a senior ranger at three dollars a day plus board and lodgings in 1938. (The appointment of a ranger from "outside" also helped quell charges of nepotism within the local families.) MacLulich had to patrol, check licences, look for poachers and serve time at Big Trout fire tower, but he did find time during his two-year stint to complete studies of varying hares, birds, and trout-attacking parasites. Rangers in those days had the power of a police constable and carried a gun when circumstances warranted. MacLulich's revolver was stolen along with his compass from his cabin on Opeongo. So few people travelled the area that it was easy to identify the thief, a young man working at Opeongo Lodge. A search warrant turned up the compass in his possession and the tyro ranger was invited to make his only arrest. Many years afterwards, the R.C.M.P. returned his revolver in good condition except for a notch filed under the barrel. MacLulich never did find out how the police recovered his weapon.[29]

The results of initial investigations of physical, chemical and biological conditions of Opeongo and other lakes in the area had surprising results, comparing unfavourably with the much richer waters of Lake Nipissing where the scientists had previously been working. With varying degrees of success, measures were initiated aimed at increasing the yield of fish from Algonquin Park lakes. Trout obtained from commercial hatcheries were planted in several lakes; minnows were introduced as food supply for the game fish; and adult trout were transferred from lightly-fished to heavily-fished lakes. A diorama in the park Visitor Centre based on decades of research at Lake Opeongo graphically explains the fish food chain.

29 *Interview with Dr. Duncan A. MacLulich at Strathroy by Ronald Pittaway, November 23, 1976.*

Throughout these formative years, all the labbers give great credit to Joe Avery and his family at Opeongo Outfitters for their friendship and practical assistance. The Averys, guides, rangers, laboratory staff and the Bates formed a close-knit community that worked well together. Joe's father, "Dad" Avery, was a guide for the Fish Lab in its early years and is remembered by Dick Miller as having a round, red, jolly face and small, lively, blue eyes. "A fringe of grey hair round a central bald spot gave him a benign, almost grandfatherly look (and) he was "short-legged, short-armed and short-necked — in fact, almost spherical." Although he was 65 years old at the time, his portaging pace left Miller far behind.

The late Ralph Bice, trapper, guide, author and Algonquin Park enthusiast, first visited Lake Opeongo in 1917 at the age of 17. He recalled in *Along the Trail in Algonquin Park* that in his early years the fishing was fabulous: "It was an easy matter to get the limit, and also easy to catch fish of ten pounds or better." Although commercial netting was allowed during World War One, Bice did not feel that this had a significant impact on the brook and lake trout population. He attributed the poorer game fishing in later years to improved access attracting more fishermen, and the introduction of bass depleting the food available for lake trout.

Fishermen tend to develop their own theories on why fish are caught–or not caught. Veteran guide Joe Lavally attributed fluctuations in large lake trout catches to changes in water levels.[30] He complained that for many years his parties caught nothing over a few pounds in Opeongo. Then the dam raised the water a few feet. "Parties went out and caught half-a-dozen 20-pounders in a few hours." He theorized that they could not have grown to that size in one season but had spent years in colder water at the very bottom of the lake. When the water level rose, he explained, it was still cold enough to suit them a few feet from the bottom–where they were much easier to catch. Scientists give no credence to Lavally's hypothesis, but that would probably not have influenced the independent-minded guide.

Black (now called smallmouth) bass were introduced with the best of intentions as noted by Park Superintendent G.W. Bartlett in his report of January 13, 1902. "The black bass put into these lakes during the past two years have increased far beyond my most extreme hopes, and are very abundant." The first report of catching a bass in Opeongo was in 1928 at Sproule Bay. Prior to the introduction of bass, the basic forage fish for lake trout was the yellow perch, eaten in shallow waters during spring and fall before the trout headed for deeper, colder waters during the summer months. Bass remained in the shallows, dining on perch and minnows all summer, leaving few for the trout. Bass also greatly depressed the crayfish

30 *At least three men with similar names were active around Lake Opeongo. Joe Lavally was the hero of Bernard Wicksteed's classic, Joe Lavally and the Paleface. As a joke on the census-taker, he had his son register as Joe Lavallee. "Hay Lake" Joe Lavalley, no relation, worked with John Bates.*

population and probably doomed the native brook (speckled) trout. This situation is true of most lakes along the Highway 60 corridor.

American eels were known to exist in Opeongo, and probably inhabited other lakes as well, but are now believed to be extinct in the park. Their demise was brought about by the construction of dams on the rivers flowing out of the park preventing either their migration to breed in the Atlantic Ocean or the return of young eels. A 27-pound pike caught in 1996 in Victoria Lake, on the Opeongo River just outside the park boundary, warned of another danger looming for the native fish. Probably one of several introduced by persons unknown and unauthorized into the lower Opeongo River during the early 1980s, this was a forerunner of pike that subsequently made their way up the Opeongo as far as Booth Lake dam. It is feared that these predators may already have circumvented this barrier, leaving only the Annie Bay dam barring their entry into Lake Opeongo from where they could appropriate the remainder of the Opeongo system.

Based on extensive studies, a variety of experiments were undertaken at the Fish Lab to adapt the park to its changing role. An alternate-year

Fish Checking Station located at Sproule Bay in 1959.
APMA 5449/MNR

lake closure system was introduced on some lakes, but not Opeongo, to give speckled and lake trout a chance to mature. No evidence of success was found so the scheme was abandoned. Because hungry lake trout are very vulnerable to winter angling, ice fishing was banned in 1955. Recognising the lure of big lake trout for the park visitor, cisco (lake herring) were introduced into Lake Opeongo as feed fish in 1948. Cisco have the same temperature and oxygen requirements as lake trout and, therefore, inhabit the same cold depths. The good news is that they eat abundant plankton which are too small for lake trout to feed on and trout thrive on cisco. The down side is that lake trout in Opeongo now mature later than they used to and about 70 percent of the fish caught is immature stock that have never spawned.

Among the continuing programs instituted by the Fish Lab was the creel census. Anglers, first on Lake Opeongo and then throughout the park, were requested to provide key information on cards distributed by summer students. The number of fish caught, location, species, size and length of time to effect the catch were recorded. With additional information assessed from stomach contents and scale samples, tabulations could be made of feeding habits, age, sex, etc. to show annual trends in any given lake. "Creel censuses" built up in this way have resulted in new management policies and widespread improvements benefiting both fish and anglers. A Fish Lab creel census diary entry for May 17, 1936 is typical of the period: "Dr. Street, Fort Erie and party of four lake trout fishing for two days. Have a cottage back of Huntsville on Sand Lake. Had good fishing, took out their limit of forty and probably ate a few also."[31]

There was early action to make the Fish Lab work relevant to the park activities. On July 17, 1937, Bill Harkness addressed the fish and game club in Pembroke. About the same time, Fred Fry and Ray Langford conducted a series of lectures on fish and aquatic biology for the young people at camps in the park. Professor Dymond did the same thing from his cottage on Smoke Lake and also established the first interpretive centre in a tent by Highway 60 at Cache Lake.

Transport at the Fish Lab improved tremendously in 1938 with the acquisition of a 1930 Ford Model A station wagon with wooden body and side curtains.

The Fish Lab has grown steadily, in both size and reputation, since 1936. Fifty scientists and support staff can be accommodated in a first-class facility; Director Mark Ridgeway emphasises the long-term nature of the research and the ability to, "put people on the front line on a sustained basis." From its first days, the Fish Lab has also been a base for non-fish

31 *Field Notes and Record Book, 1937. Evans-Campbell Listing #65, MNR Library, Peterborough.*

research; for example, Doctors Murray Fallis and Seymour Hadwen of the Ontario Research Foundation worked there on deer parasites in 1937. Of necessity, deer had to be killed for examination, and Victor Solman remembers that the venison provided a welcome variation in a rather dreary diet.

One paper serves as an example of the hundreds of scholarly works emanating from Lake Opeongo.[32] Jim Fraser, Director of the Fish Lab from 1965 to 1975, was an expert on brook trout. His work, "The Effect of Competition with Yellow Perch on the Survival and Growth of Planted Brook Trout, Splake and Rainbow Trout in a Small Ontario Lake," was judged by the American Fisheries Society to be the best paper of 1978, no mean accomplishment considering the importance and size of fisheries research in North America. What the rather ponderous title covered was a detailed analysis of fish growth in appropriately-named Little Minnow Lake, 1.5 kilometres east of Sproule Bay. The small lake originally contained only sticklebacks and four species of minnows. Beginning in 1962, the lake was stocked each year with various combinations of brook trout, splake and rainbow trout. Fishermen happily reported their catches at the Opeongo creel-checking station. Yellow perch were discovered in the lake during 1968 and Fraser seized the opportunity to investigate exactly why trout did so poorly in competition with yellow perch. Continuing his experiments for a decade, he determined that there was no particular decline in the number of trout, but their growth rate was drastically reduced while the perch prospered until their growth, too, was limited by available food. Analysis of trout stomach contents, sure enough, showed that they originally ate many large food items such as minnows, leeches, dragonfly nymphs and crayfish but, in competition with perch, had to make do with small items such as caddisfly and midge larvae. A resulting rule prohibits the use of live bait in Algonquin Park lakes. Unfortunately, the rule is almost impossible to enforce and all too often, fishermen have dumped the contents of their minnow pails, introducing perch or other trout competitors into good trout lakes.

Honours abound for work done at the Fish Lab: degrees by the dozen, but perhaps more significantly recognition by peer groups. For example, Nigel V. (Nick) Martin, who was associated with the laboratory all his professional life, received the prestigious Directors Award from The Friends of Algonquin Park in 1988. The posthumous citation reads, in part, "World-renowned as a major authority on lake trout, he was author of at least 52 papers and articles based on his fisheries work in the park, including major syntheses of work on the effects of effects of introductions and of exploitation on lake trout."

32 *Dan Strickland. "A Tale of Two Fishes." The Raven, Vol. 21, No. 6, July 24, 1980.*

Although the focus of Lake Opeongo scientific attention is on fish, this is by no means the only subject of interest to scientists. Many of the investigations under way are covered in the park's educational newsletter *The Raven*, published 12 times each year by the Ministry of Natural Resources until 1993 and subsequently by The Friends of Algonquin Park. The nesting and hunting habits of the Merlin in Algonquin Park, for example, are described by Chief Park Naturalist Dan Strickland in Vol. 26, No. 8, August 8, 1985. This small member of the falcon family can often be seen over Opeongo. The fact that Merlins never chase prey into the thick forest and must have lots of open space for a successful attack on smaller birds dictates that they generally nest and hunt near lake shores. A striking similarity was identified in that most of the Merlins observed in the park nested on islands in large lakes. It was deduced that small birds, normally smart enough to stay within the forest cover, would brave short trips across open water to greener pastures on an island–and be vulnerable to attack. Of course, several islands, separated by distances small enough to tempt crossing by small birds but open enough for Merlins to operate effectively would be an even more productive hunting ground. Strickland suggests that, if this reasoning is correct, it explains why nesting Merlins with many mouths to feed can be seen at the group of islands at the north end of the South Arm. "You may be lucky enough to see one single out a small bird, turn on the after-burner, and pluck it out of the air."

Cameron Williams, inventor of the "Williams Wabler" lure, from Fort Erie, Ontario, was a frequent visitor to Opeongo and helped the Averys to organize an annual Fish Derby. Usually held at the Dennison farm, these added a little excitement to the lives of the labbers. But, as M.J. McMurtry and B.J. Shuter recalled, "you knew you had to work like mad."[33] Every fish caught had to be checked by the lab personnel and the work was particularly intense when prizes were given for the largest, the smallest, and the most fish caught. "They entered every fish they had because they might win something." Williams provided the prizes, usually an eclectic assortment of lures, and Ken Avery's wife Juanita recalls that entry fees usually purchased equipment for the Whitney Recreational Association.

Despite a subsequent busy career as a wildlife artist, Robert Bateman has clear memories of his time catching "forage" fish by gill net and minnow trap in small tea-coloured lakes as part of the census. A conservationist at heart, he remembers: "Sadly, I was expected to dump all the fish on the ground after tabulation and leave them there to expire, so that I would not count them over again. On one occasion I came back to the site over twenty-four hours later and one four-inch bullhead was still hopefully flopping about. I rewarded him by returning him to the lake."

33 *APMA 3917 Lake Opeongo Creel Survey: Interviews with Survey Personnel, 1936-83*

Gull Island – Lake Opeongo,
1951 by Robert Bateman.
With permission.

Robert Bateman was a
student researcher at the
Fish Lab in 1951. His main
job was measuring and
taking scale samples of fish
for the creel census but he
painted in all his spare
time. He recalls, "The most
memorable painting was
done in company with Dr.
Langford who also enjoyed
getting out the brushes and
paints. We went up to the
North Arm in a red,
motorized pointer,
anchored off an island and
both went to work. I
completed my 24- by 30-
inch oil painting on the
spot. I suspect that the
rocking of the boat
contributed to some of the
rhythms in the painting."
The painting is owned by
his long-time friends,
Lorraine and Donald Smith.

There was a sequel to this
story in 1990 on the
occasion of a retirement
party for fisheries
technicians Jack Murdock
and Bernie Kukhta. Both
men were highly respected
and had completed 35
years' service; among the
many guests attending this
important event was
retired Fish Lab Assistant
Director Dr. Ray Langford.
Chatting after the cere-
mony, Langford asked the
current director, Dr. Mark
Ridgeway, if a painting that
used to hang on his cabin
was still around. It was, an
oil about 24 by 30 inches, a
forest scene in surrealistic
style, still in its original

Gull Island in the North Arm established a place in lake lore when
Bud Williams netted about four bushels of spawning whitefish many miles
away at the Annie Bay dam. Inviting fellow fishermen to a fish fry, he
concocted a story that each one had been caught off Gull Island with a
single grain of rice on a hook. In no time, the store was out of rice and
Gull Island was surrounded by frustrated fishermen being bombarded by
disturbed and angry gulls. In today's more environmentally-conscious age,
park visitors are encouraged to steer clear of the island during the nesting
season so as not to interfere with the natural production of young gulls.

Paul Shibata, manager of Fishing Buddies in Ottawa, says the
Williams Wabler lures have stood the test of time and have a huge
following. Their success is primarily due to experiments over many years
on Lake Opeongo. Sixteen-year-old Cameron Williams accompanied his
father and uncle to the Klondike a century ago, establishing the family
fortunes and future career path in gold refining and fabrication of precious
metals products at Fort Erie, Ontario. He and his son Bud, both keen
fishermen, experimented with silver-plated, spoon-shaped lures after
hearing of a giant trout snatching at a silver spoon accidentally dropped
overboard near their summer home on Kawagama Lake, southwest of

Algonquin Park, about 1932. The word got around and the Williams were soon in the fishing tackle business. The Williams were a large family and often at Lake Opeongo where they camped, usually in the East Arm, or stayed at Opeongo Lodge. Fascinated with the potential of the "spoon," Cameron and Bud continued development at Opeongo and some 200 variants of the original lure are now available. Plating one side of a spoon with gold and the other with silver, they had great success at Dickson Lake, pulling out "uncatchable" 15-pound trout. Believing the lure to be too expensive, they had no plans for manufacture until the fishing fraternity at Opeongo Lodge heard about it. They left a couple of days later with orders for 20 dozen–whatever the price.

Jack Murdock, fisheries technician from 1955-90, remembered a guide from Baysville who had lots of business on Opeongo because he supplied everything needed to fish–and guaranteed a catch. "No fish, no pay." His secret weapon was one of the earliest echo sounders. The sounder would find a school of cisco and the lake trout would not be far behind. Local resident Dave Harper has fished Lake Opeongo since 1970 and says it consistently provides "the best fishing for trout and bass anywhere."

Records of the early days at the Fish Lab and subsequent interviews clearly show that the labbers took a great deal of satisfaction from their low-budget pioneering work. They established a tradition of excellence and "getting the job done" as did amateur athletes of the same era who were described by record-breaking athlete Bruce Kidd in *The Struggle for Canadian Sport* as having "a spirit of joyous improvisation."

As is often the case with scientific research, unexpected benefits occurred in an entirely different field. With the advent of World War Two, several of the Fish Lab scientists enlisted in the RCAF. Squadron Leader Fred Fry received the British Empire Medal for his work in aviation medicine. He led a team including Flight Lieutenants Kennedy and Martin who used their knowledge of respiration to develop the Franks flying suit that assisted fighter pilots to withstand the pressures of high-gravity manoeuvres. Also among their inventions was an innovative valve which vastly improved the delivery of oxygen, vital to aircrew at high altitudes. The Fish Lab scientists were led to aviation by the force of wartime priorities, but aircraft had already made their mark on Lake Opeongo.

rather crude birch frame, still where he left it. Langford intimated that he would like to have it, to which Ridgeway without hesitation agreed, "We've enjoyed it all these years, you take it." Langford replied, "Thank you, thank you, Robert Bateman and I painted that." Bateman is not by nature a painting collaborator (he objected to working on a group mural as a child in kindergarten) and the most he contributed was probably a few brush strokes but Ridgeway was left wondering if he had given away a valuable "Bateman."

7
Aviation

JUST AS AIRCRAFT PLAYED an important–probably vital–role in the evolution of Canada, so it was with Algonquin Park. The famous "Flying Superintendent" Frank A. MacDougall flew missions through his territory in the 1930s, effectively erasing the poacher problem that had plagued the park since its inception. From his base at Cache Lake, he would fly over the vast expanse of the park and there was no escaping him in winter as tracks were plainly visible from the air. In summer, he would land alongside suspect canoes to check that fishing limits had not been exceeded. Even before MacDougall and Fred Hughes, his mechanic, arrived in 1931, however, the wide expanse of Lake Opeongo was attracting military, private, commercial and government pilots.

Canada had reluctantly accepted the "Imperial Gift" of 115 unused World War One aircraft offered by Britain as an embryo air force to each of the Dominions. The government was understandably apprehensive that this gift would incur operation and maintenance costs much greater than its value, estimated at over five million dollars. The arrival of the aircraft in 1920, however, had a fundamental effect on the development of Algonquin Park in general and on Lake Opeongo in particular.

Enthusiastic pilots of the fledgling Canadian Air Board (not RCAF until 1924), determined to prove the value of aircraft to a sceptical federal government, abandoned wartime roles in favour of fishery inspection, smuggling patrol, forest fire detection, police support, resource identification and aerial survey. None of Canada's aircraft of the period were armed. Ontario, with its vast forest and mineral resources barely even catalogued, recognised the potential offered by aircraft. The Department of Lands and Forests authorised evaluation flights in 1920 and two years later put three HS2L flying boats and supporting Avro 504 float panes on fire detection and control over three newly-created fire districts - Muskoka, Parry Sound and Algonquin Park. Fires were reported by dropping a message at a fire ranger station or by landing near a telephone or telegraph office. Notoriously unreliable engines and fragile wood and canvas airframes made the large and centrally-located Lake Opeongo a frequent haven for these pioneering aircraft.

Flight logs in Algonquin Park Museum Archives record a forced landing of HS2L G-CYAG at Sunnyside in the East Arm on September

Sixty-two of the 100 World War One aircraft given to Canada were two-seater Avro 504s, one shown here equipped with floats for operations in northern Ontario, c.1922.
JOHN GRIFFIN

In addition, 12 Curtiss HS2L flying boats were donated to Canada when the U.S. Navy departed from Dartmouth, Nova Scotia. *JOHN GRIFFIN*

Plenty of fresh air in an HS2L cockpit. District Forester W. A. Delahay on the right with aircrew (left to right) Jack O'Gorman, George Brookes and Frank Watson. Cameras were mounted in the foreground open cockpit. *JOHN GRIFFIN*

The first "Flying Superintendent," Frank MacDougall, preparing to depart from the dock at Sproule Bay in his Fairchild KR-34C. Ranger Jim Shields on the left, the man wearing a broad-brimmed hat is John Bates, man with pipe unknown.

CF-AOH was purchased new in 1931 by Ontario Provincial Air Service and assigned to Algonquin Park. MacDougall utilized it until 1938 when he relinquished it for a more powerful, enclosed-cockpit Stinson Reliant. CF-AOH was sold to private interests in 1944 and crashed at Wildcat Lake, Lake Superior Park, in 1947. It was salvaged in 1963 and rebuilt to flying condition by the Ontario Provincial Air Service for its 60th anniversary in 1984. When the Canadian Bushplane Heritage Centre established itself in the old OPAS hangar at Sault Ste Marie it "inherited" the restored aircraft which it continues to display. MNR Peterborough Historical 110-1

MacDougall, superintendent from 1931 to 1941, pioneered many applications of aviation in Algonquin Park. He left the park to become Deputy Minister of Lands and Forests, a post he held for 25 years. In recognition of his work on behalf of Ontario's parks, the section of Highway 60 through Algonquin Park was named the Frank MacDougall Parkway in 1976.

34 *Private flying was confined to 22 lakes in 1943 and further restricted in 1955 and 1973. Smoke and Kioshkokwi lakes were then the only two available and in 1987 prior permission was required to use only Smoke Lake. In 1993, all private flying into Algonquin Park was prohibited.}*

25, 1922 after delivering spare parts to a sister flying boat at Sproule Bay. Flight Lieutenant George Brookes and his crew repaired their aircraft and returned to the base at Whitney.

By 1924, the Ontario Provincial Air Service was established and had a fleet of 14 HS2Ls, one Loening flying boat and four de Havilland Cirrus Moth two-seaters. (Over the years, the fleet was modernized and expanded; by 1956, OPAS had 40 de Havilland Beavers and five Otters. Bruce West in *The Firebirds*, a history of Ontario aviation, argues that the OPAS purchases from de Havilland were the key reason for establishing the Toronto company. A total of 2,158 Beavers and Otters were eventually sold to 60 countries and de Havilland is today a world leader in the design and production of commuter aircraft.)

Early action was taken to control private aviation in Algonquin Park by threat of seizure that would make aircraft the property of His Majesty unless the regulations were complied with:

> 40. Every pilot of an aeroplane, other than those operated by the Ontario Government, shall upon making port within the Park, secure a permit from the Superintendent for which a fee of $1 per day shall be charged. An additional fee of 25 cents shall be charged for each passenger taken on flights in the aeroplane. The pilot must produce his licence whenever or wherever called upon to do so by the Superintendent, the Park Ranger or any other officers appointed by the Minister of Lands and Forests.
>
> *Regulations Respecting Algonquin Park, 1931.*

Until 1973, when private flying was banned at Opeongo, Sproule Bay was a favourite base for fishing parties.[34] A wide variety of aircraft were regular visitors. Gordon Palbiski, then owner of Hay Lake Lodge south of Whitney, flew in fishing parties one at a time to Lake Opeongo in his two-seat Piper Cub with a canoe slung alongside the floats. He would make return visits to air-drop supplies alongside fishermen who had portaged away from Opeongo to lakes where aircraft were not permitted to land. Quinte Airways with its pair of de Havilland Fox Moths initiated a service for flying fishermen from Trenton in 1946.

Among the other aerial visitors was Tommy Douglas (not the politician) of Muskoka Air Services with his Piper Super Cub. Reg Williams would fly in family and friends in his Waco biplane and, later, a Super Cub to Opeongo Lodge or to "tent city," their favourite family camping ground in the East Arm.

Despite regulations, airmen had difficulty resisting the temptation of free fishing but avoided Lake Opeongo to minimize their chances of being observed by rangers. RCAF Fairchild 51 GC-CYV, stationed at Ottawa, had recently been converted from a Model FC-2 (i.e., fuselage enlarged and strengthened) when it paid a call at nearby Happy Isle Lake in 1931. Approaching the Brice family who were camping there, the passengers borrowed a canoe and fished while the pilot taxied upwind and drifted down until enough trout were caught.
APMA 1428/W.L. BRICE

Joe Avery, proprietor of Opeongo Outfitters, on left, with camper Eddie Englehart by a de Havilland Fox Moth, c.1950.
CLOVER PALBISKI

The availability of aircraft was of great value in an emergency and several medical evacuations were carried out. In 1955 Joe Avery died while he and his wife Myra were visiting Toronto. Myra, left alone in the city, needed help. Orillia Air Services came to the rescue, flying Gordon Palbiski from his fishing expedition in Lake Lavieille to Sproule Bay where they picked up his wife Clover (Joe Avery's daughter) and her brother Alvin and flew them to Toronto Island Airport, much to Myra's relief.

OPAS de Havilland Beavers lined up for inspection at the East Arm in 1952 during the British Empire Forestry Conference. From right: George Phillips, Tom Calladine and Tom Cooke.

LOTTIE CHAPMAN/CANADIAN BUSHPLANE HERITAGE CENTRE

Many camps were organised for senior bureaucrats during the 1930-50 period, usually at the sandy beach on the East Arm and always featuring fine food and fantastic fishing. Guide George Pearson conveyed the flavour of these events during an interview by Rory McKay: ". . . somebody had some pull with a member of parliament or some high guy. They used to come in there, you take them up to the East Arm of Opeongo and leave them there, go up in every couple of days, they stayed there for a week, maybe ten days. Oh, I was running up there all the time."[35]

Typical of the more formal events at the East Arm was the British Empire Forestry Conference sponsored by the Federal Government in 1952 to enable foresters from different parts of the Commonwealth, the U.S.A., and other countries in the United Nations to share ideas. Some 50 foresters and support staff were ferried to Lake Opeongo by six Department of Lands and Forests de Havilland Beavers. They were able to savour the rugged scenery while watching demonstrations of fire fighting techniques and discussing modern forest management practices.

Main base of the Department of Lands and Forests Aviation Division was at Sault Ste. Marie and the Canadian Bushplane Heritage Centre there continues a fine tradition with the active support of current and

35 *May and George Pearson interview by Rory McKay at Bancroft, c.1975.*

Ken Avery remembers transporting about 60 air cadets from his dock to the East Arm beach in the 1950s with the help of, from left, Cadet League Director Jim Smith, Ranger Tom Murdock, Director Al Law and Ranger Charlie Levean.

Al Law organised the international air cadet exchange camp in Algonquin Park during the 1950s and '60s. The army provided a field kitchen and the park staff erected tents, provided guides and organised lectures on wilderness topics. The cadets were normally based at Mew Lake and made an expedition to Lake Opeongo where they were met by officials from Ottawa, brought in by RCAF Otters.[36] APMA 1792/Marguerite Levean.

retired personnel. As might be expected, many aviation tales are told around the coffee pot. George Beauchene was rumoured to maintain some of the habits he developed driving a tank during World War Two and he delighted in giving his passengers an exciting ride. Technician Sam English told a story of revenge that Beauchene remembers with a laugh. Beauchene took John Baulke to Opeongo to help clean up after the royal wilderness camp in 1959 but would not leave the safety of his aircraft when he spotted a snake on shore. Baulke teased him without mercy about this weakness and was threatened with some really fancy flying on the return trip. Baulke, however, carried on board a small snake and threatened to slip it down the terrified pilot's collar. "We flew about 500 feet above Opeongo and he just landed. He jumped out and never even tied the plane. He was sicker than a dog."[37] Beauchene admits that he flew straight and level but maintains that it was a BIG snake.

The departure from Sproule Bay of Noorduyn Norseman CF-AYO on August 28, 1953 with Roy Downing at the controls, two passengers and 340 kilograms of fire-fighting equipment ended in tragedy. They took off about three p.m. on a very hot day for Round Island Lake, eight kilometres to the east and one of the bases for fighting the Shirley Lake fire that consumed 538 acres of forest, the worst in a bad fire year. Ron Florent of Madawaska was manning a pump on Shirley Creek, the fire's western limit (although it did briefly invade the western bank). He vividly

36 *Personal communication with Dave Logan, Air Cadet League*
37 Interview by George Campbell c.1975. APMA.

The Norseman was the first true bush 'plane, built in Montreal for the harsh and unforgiving Canadian environment, and CF-AYO was the first one off the line in 1935. "AYO" was a film star, playing a leading role, along with Brenda Marshall and James Cagney, in the bush flying epic Captains of The Clouds. This photograph was taken on location at "Lac Vert Trading Post," built by Warner Brothers at the westerly end of Four-Mile Bay on Trout Lake, near North Bay, August 2, 1941. For reasons unknown, the registration was changed to CF-HGO for the movie.
COURTESY RUDI MAURO.

remembers the sudden roar of AYO's engine attracting his attention over the din of the fire–itself, "like two freight trains thundering through the forest"–as Downing unsuccessfully tried to climb over a hill to the south of Round Lake. A horrified Florent saw AYO slip sideways and disappear behind the hill–silence–and then smoke rose above the tree tops. Downing, Leslie Booth from Huntsville and 14-year-old Murray Bulmer from Cobden all died in the crash which, ironically, started another fire west of the Shirley Creek barrier. Already on the front line of a huge fire, Florent was more than a little concerned to have another start behind him. John Shalla and others fought what became known as The "AYO fire" but it was not officially extinguished until September 5 after it consumed almost two acres.

The wreck of AYO remained, protected by its isolation from souvenir hunters and other sticky fingers, until Bob Thomas, project manager at the Bushplane Heritage Centre in Sault Ste Marie, negotiated salvage rights. With the help of park staff and volunteers in October 1992, birches which had grown through the fuselage skeleton were removed in small sections to avoid further damage. Fuselage, floats and engine were removed to uncover, with a bit of digging, other remnants including metal dishes, fire pails, a silver fork, part of a fire hose, and corned beef and bean tins that had exploded from the intense heat of the burning aircraft. A marker was left at the site and the recovered remains are now in the Canadian Bushplane Heritage Centre, a tribute to a famous aircraft and a memorial to its last occupants.

Pilot George Campbell, now retired from the forestry department with 15,000 hours over Ontario in his log book, frequently flew into Lake Opeongo. He recalls pilot Eddie Thomas and engineer Lloyd MacKenzie arriving at Joe Avery's dock at Sproule Bay in 1953 after a long photo flight in a de Havilland Otter. They were both single: "Married men would not take the job." Joe's daughter Ellie, "a pretty, shapely girl," had her eye on the good-looking Eddie and made a few discrete enquiries of Lloyd, who tried to further his own ambitions and thwart Eddie's by hinting that his pilot was married. Ellie was pretty cool toward Eddie for a while but she soon uncovered Lloyd's subterfuge: true love triumphed and within a year Ellie and Eddie were married and moved to Sault Ste. Marie.

One cold February day in 1979, Campbell rescued two park employees, Kevin Heins and Jan Larsen, on Lake Opeongo. Keen cross-country skiers, the pair left their car on Opeongo Road at the Cameron Lake Road gate and set-off in fine weather accompanied by Heins' four-month-old puppy, appropriately named "Fifty Seven Varieties," for the Annie Bay dam. They planned to continue through the East Arm and

Lake of Bays Air Service operated two amphibious Republic Seabees, one shown here, stationed at Opeongo Lodge for the summer months of 1948-54.
APMA 1832/DARRAUGH

During the 1950s RCAF 408 Squadron, based in Ottawa, trained pilots to operate its float-equipped de Havilland Otters from a temporary base at Golden Lake, east of Algonquin Park. Lake Opeongo, with its excellent fishing and the facilities at Opeongo Outfitters was considered to be an ideal destination for training flights. MN679 is seen moored to the Opeongo Lodge dock, c.1957.
APMA 6696/MICHAEL AVERY

return via the South Arm but the trip took longer than expected and darkness fell while they were still at the north end of the South Arm. The lightly-clad men desperately sought shelter, unaware that none existed on the lake north of Sproule Bay. The puppy could not keep up so Heins made it comfortable in a snow shelter on an island. The men struggled south keeping the shoreline on their left. Almost at Sproule Bay, they did not realize Bates Island was an island, turned right along its northern shore, crossed the narrow gap to the lake's western shore and continued with the shoreline on their left–now heading north. Hoping in vain to get a bearing from the sound of the generators at the Fish Lab, which was now behind them, and exhausted, they were forced to rest in the shelter of a fallen spruce tree at Graham Bay. Frozen and with no means of making a fire, Heins was sure they would die. Taking turns keeping each other awake, they did survive the night but could barely walk the next morning, much less ski. Colleague Mike Kilby raised the alarm the following morning and set out on their trail by snowmobile at daybreak. George Campbell in his Turbo-Beaver had picked up radio calls about the missing skiers and made a pass over the lake. Imagine their relief to hear the music of a PT-6 turbo-prop approaching. In very short order, Campbell had them aboard and delivered to the East Gate and the attention of Dr. Lloyd in Whitney. Both spent several weeks in the St. Francis Memorial Hospital at Barry's Bay, and nearly lost toes to frostbite. Kilby brought back an unharmed but worried pup to visit Heins in hospital.

38 *Personal communication with Campbell and Heins*

Although not as frequently seen as in previous decades, aircraft still play their part at Lake Opeongo. Helicopters clattered overhead in 1987, transporting moose to Sproule Bay and other accessible points for shipment to Michigan. Once native to that state, moose disappeared in the 19th century but the restored habitat was once again suitable for the big animals. Algonquin Park contributed a total of 58 from its herd of about 4,000 moose in 1985 and 1987. Animals were delivered to road-access points by helicopter and loaded into specially-designed crates for the truck journey to Marquette in Michigan's upper peninsula. The December 1996 count in Michigan was 590 moose - a ten-fold increase.[39]

Photo: Sproule Bay, 1987.
JACK MIHELL

In Sproule Bay, the outfitting store and the fisheries laboratory make a connection with past events but once past Bates Island there are only minimal traces of people who passed through Lake Opeongo. Nothing can be seen of native people who hunted, fished and gathered here for centuries. Perhaps their spirits are watching, along with those of Henry Briscoe, Alexander Shirreff, Duncan McDonnell, Alexander Graham, Robert Bell, A.H. Sims, John Dennison and his family, James Dickson, Holman James, Wilbur Batchelor, Sandy Haggart, Jack Whitton, Joe Avery, Frank MacDougall, Belle and John Bates, Mr. Moodie, John Robins, George Holmberg, Aubrey Dunn, George Heintzman, Bill Harkness, Dick Miller, Nick Martin, Ralph Bice and many, many others who made their mark, by foot, by paddle or in the air. Some unfortunate travellers have disappeared under the white caps, others may be buried here. All have contributed something to make Lake Opeongo what it is today.

John Macfie, in his *Parry Sound Logging Days*, describes many conversations with men who worked in the lumbering industry, including Jim McIntosh, born in 1896, who spent his whole working life in the bush:

39 *Personal communication with Mike Wilton and Jack Mihell*

"If I was going to live life over, I'd want to do just what I did before." Jim McIntosh knew 52 songs by heart, most learned around a stove on Saturday nights in the logging camps. One recorded and transcribed by Macfie described a logging crew's journey to Lake Opeongo and is repeated here as a tribute to the men and women who have walked through these pages:

> Oh, come all you boys if you'd like to hear,
> how I got up to the bush last year.
> Unto a place you all do know,
> It's a great big lake called O-pe-on-go.
>
> (Refrain)
>
> Come a-ram-tam-tam-ta a folly-deedle-da and a roar and rum-come-away.
>
> Oh, from Arnprior we did set out,
> with old Jack Brant for to show us the route.
> We marched from there unto Renfrew,
> and there we met our gallant crew.
>
> So, to make acquaintance we all did begin,
> but some did dip too deep in the gin.
> And some jolly lads got on a spree,
> and to hire a rig they did agree.
>
> Oh, I'll say by God didn't we feel big,
> into a nickel-mounted rig.
> And into Dacre town we hoisted sails,
> and they all thought 'twas the Prince of Wales.
>
> Our teamster's name was Flanagan Coates,
> We paid well and he fed long oats.
> He said, "Come here, God-damn you Jack,
> Can't you hear that whip go clickety-clack!"
>
> Oh, Spratt came out for to welcome us in,
> and he laid down both whisky and gin.
> The landlord's treat was a merry round,
> and we drank a health to Dacre town.

So, it's now we are up to the O-pe-on-go,
 to toil all winter through the frost and the snow.
Jolly lads though we may be,
 it is in the springtime we'll get free.

Oh, the first man came with the foreman's team,
 you would swear by God they were drove by steam.
Five-trip haul on a four-mile road,
 and forty-five big logs to the load.

Oh, the next lad came with a span of blacks,
 and they were the boys you couldn't hold back.
Out in the morning feeling fine,
 and in at night way behind.

So, it's now we're through with the O-pe-on-go,
 we toiled all winter through frost and snow.
And jolly lads we still will be,
 for in the springtime now we're free.

APPENDIX 1

Opeongo Lake Chronology

1826   Lt. Briscoe and Ensign Durnford are the first Europeans known to venture through Lake Opeongo.

1829   Alexander Shirreff passes through Lake Opeongo en route from the Ottawa River to Georgian Bay.

1847   Duncan McDonnell conducts a timber survey of the Opeongo watershed.

1851   Robert Bell survey of Ottawa & Opeongo Colonization Road.
-52

1853   A.H. Sims survey of Colonization Road to The Narrows.

1860s   Squared pine logs are collected in Lake Opeongo and flushed down the Opeongo River, aided by a dam built at the foot of Annie Bay.

1871   The Dennisons establish a farm at The Narrows.

1881   Captain John Dennison is killed by a bear.

1884   The Dennisons move out. Their farm is expanded by the Fraser Lumber Company.

1885   James Dickson survey of Bower Township.

1893   Algonquin Park formally established.

1902   The St. Anthony Lumber Company builds a spur railway line from Whitney to Sproule Bay.

1916   Jack Whitton is awarded a commercial fishing licence.

1920   Col. James requests permission to establish Camp Opeongo at the Dennison Farm.

1926	Sandy Haggart obtains a Licence of Occupation to use the St. Anthony Lumber cabin at Sproule Bay.
1928	Haggart establishes Opeongo Lodge, a fishing camp.
1929	John R. Bates obtains a cottage lease in East Arm.
1930	John R. Bates lease transferred to Bates Island.
1936	Haggart sells Opeongo Lodge to Joe Avery.
	Highway 60 is completed across the park.
	The fisheries research laboratory is established in temporary quarters at Opeongo Lodge.
1937	Laboratory staff utilize a highway construction camp at Costello Lake, completing the move to Sproule Bay by 1940.
1948	The cisco, or lake herring, introduced as prey for lake trout.
1958	Ken Avery takes over Opeongo Lodge.
	Opeongo Lodge becomes Opeongo Outfitters.
1969	John Bates dies and his cottage is destroyed about 1972.
1977	The Miglin brothers take over the Opeongo Access Point Concession as Alquon Ventures Inc.
1990	The Swifts take over the Opeongo Access Point Concession as Opeongo Algonquin.

APPENDIX 2

Origin of some official and colloquial place names in and around Opeongo Lake

Annie Bay. Generally believed to be named after Captain John Dennison's daughter Ann who probably did not accompany the family to Lake Opeongo. His granddaughter Annie, daughter of Henry, did live with the family at Opeongo. Captain Dennison's wife Ann Sanderson died in Quebec in 1850. Possibly, the bay was named in honour of them all. It was previously called Graham Bay for Alexander Graham of Renfrew who cut and squared pine in the area during the 1870s, about the time the Dennisons arrived. On an Ontario Department of Lands and Forests map of Algonquin Provincial Park c.1905 by E.H. Harcourt (OA B-43) it is called South East Bay.

Bates Island. Named after John R. Bates, Packard dealer from Youngstown, Pennsylvania, the only person to build a private cottage on Lake Opeongo.

Blueberry Island. Named by 36th Hamilton Boy Scout Troop that camped here for several years after World War Two.

Dead Man's Bay. A truck belonging to the J.R. Booth Company went through the ice with three men in the cab in 1938. The two sitting on the outside managed to escape.

Depot Farm. After the Dennison's left about 1882, their farm was expanded by logging companies and maps of the period identified it as Depot Farm.

Englehart Island. Eddie Englehart had a tent camp here during the 1940s and 1950s. The adjacent small island is Little Englehart Island.

Fishgut Bay. Unromantically named as the depository of entrails from Fish Lab investigations.

Gull Island. A well-used nesting site for Herring Gulls. Also known as Hershey Island.

Graham Bay and Creek. Named for Alexander Graham who cut pine around Lake Opeongo in the 1870s.

Hailstorm Lake and Creek. The name was probably given to the lake by James Dickson during his survey of McLaughlin Township in 1883. Possibly, his survey party was caught in a hailstorm. He named the creek Black Creek because of the dark colour of its water which seeps through an immense bog, noted for its wildlife.

Happy Isle Lake. Timber cruisers called it Johnson's Lake but Dickson identified it as Green Lake (from its colour) in 1885. The name was changed to Happy Isle at the request of Mrs. Van Ness Delamere whose husband and one son were drowned near their favourite camping spot on "happy isle" during a severe wind storm in August 1931. A plaque remembers the event: "One wild and awful moment and then – God." The creek took its name from the lake.

Hartley Lake. Named for the son of Ranger George Holmberg. Hartley, a Fish Lab technician, died of injuries sustained in a motorcycle accident on Opeongo Road in 1965.

Hurricane Island. A bud worm infestation killed many balsam trees on the east side of the island about 1968. The trees soon rotted and fell, giving the appearance of being blown down by a fierce wind. Tangled branches make this part of the island almost impossible to traverse on foot. Also known as Bannock Island.

Langford Lake. South of Hailstorm Bay, named after Professor Ray Langford, one of the original staff members of the Fisheries Research Laboratory.

Little Minnow Lake. East of Sproule Bay, it originally contained only sticklebacks and four species of minnows and was the site of important research on the growth of trout in competition with yellow perch.

Lucky Strike Point. A rocky shoal inhabited by cisco, prey of lake trout, and a spawning ground for lake trout.

Myra Lake. Nestled in the hills west of Sproule Bay, it is named for Myra Avery, wife of Opeongo Lodge owner Joseph Avery.

Opeongo. Believed to be derived from the Algonkin "Ope au wingauk" translated as "sandy at the narrows." This description likely refers to the sandy lake bottom at The Narrows. However, Michael Bernard, in a 1977 interview by Ronald Pittaway at Golden Lake Reserve, explained that Indians seldom named a lake per se, but conveyed a meaning or location. "Opeowongon," for example, meant a sandy area bordering the water, possibly identifying the sandy beach in the East Arm.

"Opeongo" was widely used. Sproule Lake used to be called Little Opeongo Lake and Sproule Creek was Little Opeongo Creek. Aylen Lake, just outside the park, was also called Little Opeongo until about 1854 but the old name persisted for many more years (Henry Dennison, son of Captain Dennison, moved here via Combermere when the family left its farm at Lake Opeongo.)

Sand Island has a sandy beach and is also known as Bear Island.

Sardine Point. Tradition has it that here small fish are more likely to be caught than large ones.

Secret Lake. West of Hailstorm Creek in North Arm and difficult to see in low-lying terrain. It was secretly stocked with speckled trout by John Bates.

Sproule Bay, Sproule Lake and the creek connecting them take their names from the township in which they are located. The township was named, in 1889, for Charles H. Sproule, Provincial Auditor, but it was never divided into concessions and lots.

Squaw Island. Identified on Department of Surveys 1928 Map of Algonquin National Park (APMA Q1.4.7) in pencil, probably by Audrey Saunders during interviews in the 1940s.

Sunnyside. The peninsula on which Henry Dennison settled was named Sunnyside, possibly because of its sun-drenched south shore, and to distinguish it from John Dennison's relatively-shady farm on the east side of The Narrows. Timbers from Henry's farm were utilized to build a ranger shelter hut, known as Sunnyside Cabin, on the south shore.

The Narrows. Connection from the South to the East Arm, this offered good fishing and a possible village, mill and bridge site serving traffic along a proposed colonization road from the Ottawa River.

Windy Point. Notorious for sudden changes in wind and wave conditions and the site of several drownings.

Wolf Island. Ranger George Heitzman deposited the bodies of wolves killed during the winter on a small island near his cabin in Sproule Bay. Malcolm Williams of the Williams Wabler family was a frequent visitor and remembers boyhood treasure hunts for wolves' teeth. (In the mistaken belief that fewer wolves meant more deer, rangers poisoned (until 1920s), trapped, snared and shot wolves at every opportunity until 1957 when a wolf research program was initiated.)

Bibliography

"A Guide to Angling in Algonquin Park." Toronto: Department of Lands and Forests, 1958. APMA.

"A Working Paper on The History of Algonquin Park, 1893 to the Twentieth Century, March 1970." The Algonquin Park Task Force History Project. APMA.

Addison, Ottelyn. *Early Days in Algonquin Park*. Toronto: McGraw-Hill Ryerson, 1974.

Addison, Ottelyn, and Elizabeth Harwood. *Tom Thomson: The Algonquin Years*. Toronto: Ryerson, 1967. Reprinted Toronto: McGraw-Hill, 1988.

Bartlett, G.W. "Algonquin Park Superintendent's Report from Mowat to Commissioner for Crown Lands, January 13, 1902." APMA 4085.

Beagan, R.S. "Captain John Dennison Saga – The Algonquin Park Settler." *Wigwam*, 4 (7), December 1978. Toronto: Ministry of Natural Resources publication. APMA 518.

Bice, Ralph. *Along The Trail In Algonquin Park*. Toronto: Consolidated Amethyst Communications, 1980.

Briscoe, Henry. "Report of a Survey Undertaken in Pursuance of the General Orders of the 18th and 20th February and I July 1826 to Examine the Water Communication Between Lake Simcoe and the River Ottawa." PAC MS c248 pp 81-85.

Census of Nipissing District (Bower, Dickson, Sproule and Preston Townships), 1881.

Christie, W.J. "The Bass Fisheries of Lake Ontario." M.A. Thesis, University of Toronto, 1957. APMA 2873.

Dill, Charles W. *Wings Over Ontario*. Toronto: Department of Lands and Forests. Reprint from Canadian Geographical Journal, February 1956.

Dunne, Aubrey. "Diaries, 1937-41." APMA

Garland, G.D., compiler. *Glimpses of Algonquin: Thirty Personal Impressions From Earliest Times to the Present*. Whitney (Ont.): Friends of Algonquin Park, 1989.

Garland, G.D., *Names of Algonquin. Stories Behind the Lake and Place Names of Algonquin Provincial Park*. Algonquin Park Technical Bulletin No 10, 1993. Whitney (Ont.): Friends of Algonquin Park.

Hotson, Fred W. *The de Havilland Canada Story*. Toronto: Canav Books, 1983.

Hurley W.M., and I.T. Kenyon. " Algonquin Park Archeology, 1970." Research Report No. 3, Department of Anthropology, University of Toronto, 1971. APMA 2476.

Hurley W.M., I.T. Kenyon, F.W. Lange, and B.M. Mitchell. "Algonquin Park Archeology, 1971." Anthropological Series No 10, Department of Anthropology, University of Toronto, 1972. APMA 2477.

Kates, Joanne. *Exploring Algonquin Park*. Vancouver: Douglas & McIntyre, 1992.

Kiems, Beverly Rae. Packard, *A History of the Packard Motor Car and the Company*. Toronto: Clarke, Irwin 1978.

Lambert, Richard S. and Paul Pross. *Renewing Nature's Wealth*. Toronto: Ontario Department of Lands and Forests, 1967.

Lundell, Liz., and Donald Stanfield. *Algonquin – The Park and its People*. Toronto: McClelland & Stewart, 1993.

Macfie, John. *Parry Sound Logging Days*. Erin: Toronto: Stoddart Publishing, 1992.

MacKay, Roderick, and William Reynolds. *Algonquin*. Erin: Boston Mills Press, 1993.

MacKay, Roderick. *A Chronology of Algonquin Park*, Algonquin Park Technical Bulletin No.8. Whitney (Ont.): Friends of Algonquin Park, 1993.

Martin, N.V. *The Harkness Laboratory of Fisheries Research*. Toronto: Department of Lands and Forests, 1968. APMA 0276.

McDonell, D. Survey Diaries, Field Notes and Reports, Madawaska River, From 5 April 1847 to 29 November 1847. AO MS924 Reel 13.

McMurtry, M.J., and B.J. Shuter (compilers and editors). *Lake Opeongo Creel Survey: Interviews With Survey Personnel, 1936-83*. APMA 3917.

Miller, Richard B. *A Cool Curving World*. Toronto: Longmans, 1962.

Murray, Florence B. *Muskoka and Haliburton 1615-1875, A Collection of Documents*. Toronto: The Champlain Society, 1963.

Pigeon, Mary McCormick. *Living at Cache Lake Algonquin Park, 1936-1950*. Whitney: The Friends of Algonquin Park, 1995.

"Photo Story – Forestry Conference." *Sylva*, Vol 8, No. 5, Sept-Oct 1952. Toronto: Department of Lands and Forests.

Quinsey, Wm. John. (compiler and editor) *Research Index of The Early Days of Land Surveys in Canada*. Scarborough (Ont): The Association of Ontario Land Surveyors, 1992.

Rayburn, Alan. *Place Names in Ontario*. Toronto: University of Toronto Press, 1997.

Robins, John D. *The Incomplete Anglers*. Toronto: Collins, 1943.

Saunders, Audrey. *Algonquin Story*. Toronto: [Ontario] Department of Lands and Forests, 1946.

Shaw, S. Bernard. *The Opeongo, Dreams, Despair and Deliverance*. Burnstown: General Store Publishing House, 1994.

Shirreff, Alexander. "Topographical Notices of the Country Lying Between the Mouth of the Rideau and Penetanguishene on Lake Huron." Transactions of the Literary and Historical Society of Quebec.11 (1831): 243-310. National Library of Canada.

Sims, A.H. "Field Note Book of The Survey of the Ottawa and Opeongo Road, 1853." MNR FNB 2230.

Strickland, Dan. "A Tale of Two Fishes." *The Raven*, Vol. 21, No. 6, July 24, 1980. Ontario Ministry of Natural Resources.

Strickland, Dan. *Algonquin Logging Museum, Logging History in Algonquin Provincial Park*. Whitney: The Friends of Algonquin Park, 1993.

Strickland, Dan. *Fishing in Algonquin Provincial Park*. Whitney: Friends of Algonquin Park, 1995.

Strickland, Dan. "Opeongo Chronicle." *The Raven*, Vol. 23, No 4, July 15, 1982. Ontario Ministry of Natural Resources.

Strickland, Dan. "The Island-Hopper-Killers." *The Raven*, Vol. 26, No. 8, August 8, 1985. Ontario Ministry of Natural Resources.

Strickland, Dan. "What Can We Learn?" *The Raven*, Vol. No.1, April 23, 1992. Ontario Ministry of Natural Resources.

Taylor, James Robinson. "The Autobiography of James Robinson Taylor, September 1963." Unpublished. Arnprior: Hank Legris.

Taylor, Robert J. "Logging in the Valley – 75 Years Ago," *Your Forests*, Vol. 8, No. 3, Winter 1975, pp 8-12. Ontario Ministry of Natural Resources.

Timber Licences. Archives of Ontario RG1 F-1-2.

Timber Limit Applications Books. Archives of Ontario RG1 F-1-4 Vols 2-7.

Tozer, Ron and Nancy Checko. *Algonquin Park Bibliography*: Algonquin Park Technical Bulletin No.12. Whitney: The Friends of Algonquin Park, 1996.

Unitt's Bottle Book and Price Guide. Peterborough (Ont.): Clock House Publications, 1985.

West, Bruce. *The Firebirds*. Toronto: Ministry of Natural Resources, 1974.

Westhouse, Brian D. *Whitney. St. Anthony's Mill Town on Booth's Railway*. Whitney: The Friends of Algonquin Park and The Township of Airy, 1995.

Wicksteed, Bernard. *Joe Lavally and the Paleface*. Toronto: Collins, 1948.

Wilson, James. *Report on The Algonquin National Park of Ontario For The Year 1893*. APMA1148.

Wilton M. "Death of Two Campers - Bates Island, Lake Opeongo, November 5, 1991." APMA.

Winearls, Joan. *Mapping Upper Canada 1780-1867*. University of Toronto Press, 1991.

Oral History Interviews APMA:

Autayo, Harry. At Barry's Bay by Ronald Pittaway, November 23, 1979.

Bernard, Michael Joseph. At Golden Lake Reserve by Ronald Pittaway, January 27, 1977.

Bolt, Blake. At Algonquin Park by Ronald Pittaway, January 3, 1980.

Clarke, Dr. C.H.D. By Ronald Pittaway, December 2, 1975.

Dunne, Aubrey. At his home in Huntsville by Roderick MacKay, November 22, 1975.

English, Sam. At his home north of the Muskoka River by George Campbell, c.1975.

Lisk, Bernice. By Ronald Pittaway, July 25, 1979.

MacLulich, Dr. Duncan A. At his home in Strathroy by Ronald Pittaway, November 23, 1976.

Martin, Dr. Nick. At Harkness Fisheries Research Laboratory by Ronald Pittaway, October 29, 1975.

Martin, Dr. Nick. At Maple by Roderick MacKay, March 1976

Nicholas, Leo and Mrs. At their home in Whitney by Joanne Dufresne, July 17, 1979.

Pearson, May and George. At Bancroft by Roderick McKay, c.1975.

Maps:

Sketch shewing modes of communication between Lake Simcoe and the Ottawa. Report of Henry Briscoe, Lt. R.E., 1826. NMC 2845.

Map of the Territory between the Ottawas River & Lake Huron . . . By Alexander Shirreff Esq. Compiled by Willm Henderson, 1831. NMC 2856.

Map by Maria Knowles Showing The "Tract of Land Proposed To Be Settled By Land Companies, 1834. NMC 10131.

Plan and Survey of the River Madawaska – a tributary of the Ottawa – beyond the Surveyed Lands commencing at the Southwesterly boundary of the Township of Blithfield under Instructions bearing date 19th January 1847. [Sgd] Duncan McDonell, Greenfields Dy Provl Surveyor. AO SR11070 No.46.

Map of the Country between Ottawa and St. Lawrence Rivers shewing Projected means of Communication. Crown Lands Department Toronto 6th Decr 1850. NMC 2884.

Government Map of part of Huron and Ottawa Territories 1863-64. Thomas Devine, Department of Crown Lands. AO B-60(1396).

Plan of Petewawa River shewing timber licences. A.J. Russel, Department of Crown Lands, Quebec 6 May 1865. AO C-61.

Plan of Upper Waters of Madawaska From Eyre and Clyde Townships northerly to Petawawa, 1871. AO R-M(U).

Timber Map Bower Township (c.1885). James Dickson. AO RG1(C66-67).

Plan of the Township of Bower Nipissing District, May 29, 1885 (sgd) James Dickson, P.L.S. CLS 633.

Plan of the Township of Dickson. Toronto: Department of Lands and Forests, 1953, replacing Survey by Thomas Byrne, P.L.S., dated October 11, 1888. CLS 893.

Algonquin Provincial Park. Department of Lands and Forests, E.H. Harcourt, c.1905. AO B-43.

Drive Route of logs from Timber Limits of McLaughlin Bros on Petewawa, Geo H. Johnson, 1921. AO B-35.

Plan sewing Part of Lot 27 Con II, Parts of Lots 27 & 28 Con III And Parts of Lots 27 & 27 Con IV, Township of Bower, District of Nipissing. Speight & van Nostrand, O.L. Surveyors. Toronto, 14 October, 1921. CLS 39342.

Map of Algonquin Provincial (Ontario) Park, Canadian National Railways, 1922. APMA Q1.4.3.

Map of Algonquin Provincial (Ontario) Park, Canadian National Railways, 1926. APMA Q1.4.4.

Map of Algonquin National Park, Department of Surveys, 1928. APMA Q1.4.7.

Plan and Field Notes. Parcels on Island "A" Between South Arm and Sproule Bay, Opeongo Lake, Township of Sproule, District of Nipissing, Algonquin Park. Pembroke, Ont Nov 15th 1930. F.W. Beatty, Ontario Land Surveyor. CLS 86054.

Plan of Part of the Point at the Southerly End of Sproule Bay, South Arm of Opeongo Lake. DLF (Sgd) Gilbert E. Ward. January 6, 1953. Date of Survey September 1952. CLS 80626.

Timber Licences, Algonquin Park. Map 47A. Department of Lands and Forests, December 1969. AO B-43.

Ontario Base Map 1017 7100 50550. 1988 MNR.

Newspapers:

Canadian Lumberman, Vol.X No.52, January 27, 1904 and Vol.XI No. 4, November 30, 1904.

The Forester, Huntsville, December 6, 1978.

The Globe and Mail, September 23, 1943.

Toronto Star, July 27, 1959.

Index

Addison, Ottelyn, 63
Aikins, Nicholas, 57
Air Cadet League, 89
Algonquin Outfitters, 38
Algonquins, 2
Alquon Ventures Inc., 37
American eels
Arnason, Jon, 70
Avery, Clover, 36
Avery, Ellie, 91
Avery, Joseph, 25, 32-36, 66, 75, 87
Avery, Juanita, 79
Avery, Kenneth, 36, 89
Avery, Myra, 36, 87
Avro 504 float plane, 84
Bartlett, G.W., 75
Batchelor, Wilbur Commodore, 45, 47
Bateman, Robert, 79-80
Bates, Belle, 50, 53, 54
Bates, John R., 39, 45-54, 75, 85
Bates, Sarah, 54-55
Baulke, John, 89
Beatty, F.W., 48-49
Beauchene, George, 89
Bell, Robert, 6, 9-10
Betteridge, Greg, 27
Bice, Ralph, 13, 29, 75
Black ash, 18
Black bears, 56-58
Booth, J.R., 13, 29, 34, 35, 39, 43
Booth, Leslie, 90
Bordowitz, Mrs., 54
Bordowitz, Peter, 52, 54
Bowers, Ned, 64
Boyle, John, 47

Brice family, 26, 43, 51, 87
Briscoe, Henry, 2-3
British Empire Forestry Conference, 88
Brookes, George, 85-86
Brûlé, Etienne, 2
Bulmer, Murray, 90
Cache Lake, 15
Cadge crib, 12-13
Cagney, James, 90
Cain, W.C., 29, 47-49
Camp Opeongo, 43-45
Campbell, George, 91, 92
Canada Atlantic Railway, 15-16
Canadian Bushplane Heritage Centre, 88, 90-91
Canadian National Railway, 43, 45
Canoe Lake, 15
Champlain, Samuel de, 2
Cisco, 77
Combermere, 19-20
Coons, Boni, 41
Costello Lake Picnic Ground, 65, 66
Coventry, Alan, 70
Creel census, 77
Curtiss HS2L flying boat, 84-85
De Havilland Beaver, 88
De Havilland Otter, 92
Dennis Canadian Company, 16, 43
Dennison, Captain John, 13, 19-22
Dennison family, 6, 19-24
Depot Farm, 23
Depot Lake, 13
Dickson, James, 1, 16, 24
Doan, Kenneth, 66, 70, 72, 73
Donly, Rea H., 40-41

Downing, Roy, 89
Drownings, 40-41
Dubé, Captain, 62
Dunne, Aubrey, 13, 40, 55, 60-61, 71
Durnford, Ensign, 2
Dymond, J.R., 64
Elton, Charles, 70-71
Engelheart, Eddie, 55, 87
Enzor, Donald, 40
Fairchild KR-34C float plane, 85
Fairchild 51 float plane, 87
Findlayson, Mort, 13, 20-21, 25
Fisheries Research Laboratory, 31, 33,
    64-78, 81
Florent, Ron, 89-90
Fraser and McCoshen, 13, 23-24
Fraser, Jim, 78
Frehe, Carola, 56
Fry, Fred, 65-66, 73, 81
Fry, Irene, 65-66, 71
George, Anita, 21
Giles, Gwen, 58
Graham, Alexander, 12
Grand Trunk Railway, 15, 41
Haggart, Sandy, 29-33, 47, 49
Harkness, Martha, 66, 68
Harkness, William John Knox, 33, 65, 73
Harper, David, 81
Hayes, Michelle, 57
Heins, Kevin, 91
Heintzman, George, 29, 49, 51, 55, 58,
    59-61, 71
Henderson, William, 5-6
Hildebrand, Mike, 57
Holmberg, George, 29, 59, 60
Holmberg, Hartley, 60
Hudson, John, 20-21
Hudson's Bay Company, 11
Hughes, Fred, 83
Hurley, W.M., 1
Huron Tract, 2-4, 7-8, 22, 23
Ide, F.P., 64, 73
Imperial Gift, 83-84
Inglis, Jeremy, 57-58
Jakubauskas, Raymond, 56
James, Holman, 25, 41-46
Kennedy, William, 66, 73, 81
Kidd, Bruce, 81
Kilby, Mike, 92

Kirkwood, Alexander, 39
Koulas, Mike, 67
Kuiack, Frank, 37
Kukhta, Bernie, 80
LaBarre, Claude, 45
Langford, Ray, 66, 69, 80-81
Larsen, Jan, 91
Lavallee, Joe, 40, 62
Lavally, Joe, 75
Law, Al, 89
Levean, Charlie, 62, 89
Linklater, Tom, 55
Lisk, Bernice, 21, 23-24
Little Opeongo Lake, 8, 23
Logging licences, 12
Lynch, Simon, 59
MacDonald, J.H., 49
MacDougall, Frank A., 33, 53, 60, 61, 64,
    67-68, 73, 74, 83, 85-86
Macfie, John, 93-94
MacKenzie, Lloyd, 91
MacLulich, Duncan A., 74
Marshall, Brenda, 90
Martin, Chip, 53-54
Martin, Nick V., 14, 22, 53-54, 59, 61, 78
Martin, Robert, 66, 70, 81
McCormick, Tom, 59
McDonell, Alexander, 12
McDonell, Collin, 12
McDonell, Duncan, 6-7, 12
McDougall, Alexander, 9
McIntosh, Jim, 93-94
McLachlin Brothers, 16-17
McRae Lumber Company, 18
Meadows, Timothy, 40
Mendelson Joe, 37
Merlin, 79
Miglin brothers, 37
Mihell, Jack, 56
Millar, John W., 29, 47
Miller, Richard B., 65-66, 71, 72-73, 75
Moodie, Mr., 33
Moose, 93
Munn Lumber Company, 15-16
Murdock, Jack, 80, 81
Murdock, Tom, 89
Nagashima, Diak, 40
New York Press Party, 27
Noorduyn Norseman, 89-91

Northcote, Jim, 37
O'Gorman, Jack, 85
Ontario Provincial Air Service, 86, 88
Opeongo Algonquin, 38
Opeongo Lodge, 25, 29-36, 64, 91
Opeongo Outfitters, 36, 75
Opeongo River dam, 13-14
Ottawa and Opeongo Colonization Road,
    6-10, 13, 19
Ottawa, Arnprior & Parry Sound Railway,
    13, 39
Oxen, 14
Palbiski, Gordon, 49, 86
Parks, Russel, 34
Patterson, Pat, 40
Pearson, George, 88
Pearson, George, 62
Penetanguishene, 2
Pigeon, Mary McCormick, 32-33
Queen Elizabeth, 61
Republic Seabee amphibian, 91
Ricker, W.E., 64
Ridgeway, Mark, 77, 80-81
Robins, John D., 55-56
Robinson, Mark, 44
Saunders, Audrey, 13, 20
Sawyer, Basil, 35
Schmanda, Jerry, 38
Schnitter, Charles, 40
Secret Lake, 52
Shalla, John, 90
Shields, Jim, 34, 85
Shirreff, Alexander, 4-7, 11, 21
Shirreff, Charles, 3
Sims, A.H., 6-10
Smith, Jim, 89
Solman, Victor, 66, 69-73, 78
Sprules, William, 68, 72
St. Anthony Lumber Company, 15, 29-
    30, 31
Strickland, Dan, 57, 79
Sultmanis, Erik, 37
Sunnyside, 25-26, 84
Swift, William, 38
Tate, Joe, 61
Taylor, James R., 17-18
Taylor, Robert J., 17-18
Telephone system, 35
Tennisco, Joe, 59

The Raven, 79
Thomas, Eddie, 91
Thomas, Robert, 90
Thomson, Peter, 63
Timber cruiser, 11-12
Torrens, Millicent, 53
Torrens, Robert G.,26, 55
Tuvi, Harry, 22
University of Toronto, 64-65
Watson, Frank, 85
West, Bruce, 86
Whitney and Opeongo Railway, 15-16, 39
Whitton, John Edward, 29, 63
Williams (Wabler) family, 61, 79, 80-81,
    86
Wilson, James, 15, 63
Wilton, Mike, 57
Wright, Fred, 29
Wyatt, Kenneth, xi
Wyatt, Thelma, xi

About the Author

S. Bernard Shaw is no stranger to readers of Algonquin Park and Ottawa Valley historical lore. In his first book, *The Opeongo: Dreams Despair and Deliverance* (1994), he traced the old Opeongo colonization road from the Ottawa River westward toward the Algonquin highlands and Lake Opeongo. He followed that with his comprehensive account of a famous Algonquin lake in *Canoe Lake, Algonquin Park: Tom Thomson and Other Mysteries* (1996). He has also written more than 200 magazine articles on historical, aviation, and business topics.

A native of Derbyshire, England, the author emigrated to Canada with his family in 1957 and currently resides near Ottawa. During a varied career he has served in the Royal Air Force; designed and developed dairy equipment; designed subsystems for automobiles, aircraft engines and research rockets; and been a business development officer with the federal government. For the past decade, his business consultancy has been increasingly neglected in favour of historical research and the publication of heritage stories which he hopes will help illuminate the achievements of pioneers and improve the chances of their patriotic dreams being realized in a united Canada.